WILL SUSTAINABILITY FLY?

T0174173

Will Sustainability Fly?
Aviation Fuel Options in a Low-Carbon World

WALTER J. PALMER

Routledge
Taylor & Francis Group

LONDON AND NEW YORK

First published 2015 by Ashgate Publishing

2 Park Square, Milton Park, Abingdon, Oxfordshire OX14 4RN
52 Vanderbilt Avenue, New York, NY 10017

Routledge is an imprint of the Taylor & Francis Group, an informa business

First issued in paperback 2020

British Library Cataloguing in Publication Data
A catalogue record for this book is available from the British Library

The Library of Congress has cataloged the printed edition as follows:
Palmer, Walter J., 1948- author.
  Will sustainability fly? : aviation fuel options in a low-carbon world / by Walter J. Palmer.
    pages cm
  Includes bibliographical references and index.
  ISBN 978-1-4094-3091-9 (hardback)
  1. Aircraft exhaust emissions--Environmental aspects. 2. Envi-ronmental policy.
  3. Airplanes--Fuel--Government policy. 4. Aeronautics, Commercial--Energy
  conservation--Government policy. 5. Fuel switching--Government policy.
  6. Sustainable engineering--Government policy. I. Title.
  TD195.A27P35 2015
  629.134'3510286--dc23

                                                                          2014028904

ISBN 978-1-4094-3091-9 (hbk)
ISBN 978-0-367-67003-0 (pbk)

Contents

List of Figures

Preface

Who should read this book? An 'aviation book' should be for 'aviation types', no? But this is not a book that simply pertains to a certain industry. It is more an examination of how an industry will confront a challenge that confronts us all. In addition, many of us use the services of that industry. Any person who has any interest at all in either flying or sustainability will need to know how aviation can address the most prominent part of its environmental and social effect in the world. Airline executives, aviation policy wonks, and academics will certainly be interested in what is written here. Presumably, they will be more interested than most. But anyone who considers their own efforts at sustainability or those of their organization will want to know about the direction being taken by an industry that supplies them an expensive carbon-intensive service. We all have enough of an idea about how sustainability might work to be able to say that commercial flight must be a big deal in that context.

Every book could be fatter. 'Sustainability' and 'commercial air travel' are both immense topics. Sustainability, particularly, is the story of everything about the human project. So the first thing to know about this book (and as the subtitle makes clear) is that, in terms of sustainability efforts, we will tackle only one particular thing: commercial aviation flight energy, or fuel. But if the focus here is on just one thing, there are still many aspects to it. Each chapter could be its own book—or ten. So the treatment of each aspect of the fuel challenge is necessarily general. My hope is that accessibility will be enhanced and much that has rested within the purview of the experts will be available to readers who, even if they are *themselves* experts in one area, will be general interest readers in others. The broader audience—those who are interested in aviation and sustainability as users of the service—will be able to read everything here.

This is not really a 'read-and-do' book; it is a 'read-and-think' book. In that regard, each chapter will unveil a whole new topic. So it may seem as if the fuel subject and its treatment are getting larger and more complicated as the book progresses, eventually becoming infinitely complex—that any specific way of addressing our sustainability challenge will elude us, receding and dropping below an ever-more-distant horizon. The very important thing to remember while reading is that in our journey we will realize not only that the solutions are at hand, but that the solutions are in action in some places, and that, in some ways, the solutions are a lot more straightforward than seemed possible as we progressed through the earlier parts of describing the challenge. We will see that there are two kinds of solutions: There are certainly technical solutions to aviation's sustainable fuel need. But then there is also the other job: Making the immense topic of sustainable fuel's possible future comprehensible to users and providers of both air service and the fuel itself. The policy establishments must enable more action, and we all need to know that such a thing is possible, and why. It is the second part—finding the political and economic way forward—that remains open.

What is written here is certainly advocacy, but it is not mere advocacy for an industry. It presumes that because commercial aviation is important, it is here to stay. We can, as providers, regulators, and users, see aviation grow in a way that somewhat ignores environmental and social matters, *or* we can imagine a different path. The commentary in these pages advocates for the latter. There are those who would contend that air transport should be reduced or even eliminated. The onus is on them to explain how that would happen.

We could say that there are two 'sides' to the story about whether commercial flight can and should be rendered more sustainable, but that would not reflect the reality that sustainability itself has a limitless number of sides. We could frame this as a bipolar contest, with airlines and aircraft manufacturers on one side and environmentalists on the other, but that would have no resonance with those for whom both the environment *and* the services that air commerce make possible are important—even essential. As one who actually lives in both camps, the approach that I have taken is to drop the whole idea of 'sides' and to simply explore the question and to try and figure out what must happen to satisfy everyone. Nevertheless, to some, it may seem like industry advocacy to merely ask the question, 'Assuming that it is possible to secure supplies of sustainable fuel for commercial aviation, how could that be achieved?' I regret that. I have spent decades flying airplanes happily and profitably. On the other hand I have invested enormous amounts of time trying to advance goals of sustainability and action on climate change. I really do favour both of these perfectly valid interests.

The goal here is to broaden and contextualize the knowledge resource concerning aviation, fuel, sustainability, and policy in a way that allows anyone to gain a more integrated understanding of the challenges that we face in dealing with both flight emissions of greenhouse gases (GHG) *and* the broader sustainability questions that attend attempts to eliminate them. Again, while this work is certainly addressed to academics, policy-makers, and air-industry leaders and stakeholders, it is my sincere hope that since flight represents a sustainability hurdle for people who fly as passengers or ship goods by air, the audience will be quite broad.

The reader will become aware that I cite online sources frequently. While much information resides in peer-reviewed publications, much does not. Also, some of the most valuable insight comes straight from the heads of knowledgeable people through interviews. Everything related to global warming, climate change, emissions reduction, and sustainability—but particularly as they pertain to aviation—is in daily flux at a frenetic pace. For this book to have relevance at time of publication, it must make reference to the most current discourse and events that afford us the views of all players. This can only be accomplished by tapping into the real-time dialogue. Also, since public perception is key to my treatment of a subject that must be comprehended as a somewhat politically driven phenomenon, with regard to desirable policy outcomes, the state of discussion in popular media is extremely relevant.

The nature of much of my research for this project was to consult broadly with those who are experts in a range of particulars. It is absolutely essential to point out that this book, as an extended discussion, relies critically upon the information and insights that such people provided. I do acknowledge the assistance that has been provided by all those whose expertise and ideas informed much of what you will read on the pages that follow. Beyond that, many others, whose skills and knowledge have virtually nothing to do with aviation, fuel, or other principal topics, also assisted me. In condensing, summarizing, and interpreting what I have learned from all of these people, and in involving them in the project, it may be that errors have occurred. Any such failures are mine and should not be attributed to those who tried to help.

Walter J. Palmer
June 1, 2014

Acknowledgements

Certain individuals made a particular difference in my ability to carry out this project.

First of all, let me express my deep gratitude to Deanna Cowan who, having volunteered for the 'lulz', ended up being my absolutely key resource in editing and research. Thank her for anything that you happen to appreciate about this book. Blame me for everything else. Many of the words and references, and much about the order in which they appear was 'suggested' by Deanna. In hindsight, I can't imagine how I thought I would get the thing done without the kind of help she was able to provide. Her efforts were valiant and critical—and there was actually little in the way of lulz, as it turned out.

My thanks to Guy Loft for expressing interest in the project and for patience above and beyond.

For granting interviews: Barbara Bramble, Linden Coppell, Paul Dickinson, Fred Eychenne, Ryan Faucett, Emma Harvey, John Heimlich, Tim Johnson, Dirk Kronemeijer, Chris Lyle, Chris Malins, Lourdes Maurice, Jonathan Pardoe, Paul Steele, Chris Turner, Jennifer Holmgren, Mike Lu, Jim MacNeill, Myrka Manzo, John Plaza, Jonathon Porritt, Eric Reguly, Pedro Scorza, Nancy Young.

For providing advice, counsel, or other kinds of help and opportunity along the way: Richard Altman, Jody Andruszkiewicz, Michel Baljet, Doug Bastow, Paul Bogers, Peter Bombay, Jude Chillman, Jenny Clad, Sabrina Cowden, Steven Davis-Mendelow, Audrey Depault, John Desrameaux, Dan Evans, Philippe Fonta, Harriet Friedman, Henk de Graauw, Al Gore, Jim Harris, John Hawksworth, James Hileman, Jo Howes, Jane Hupe, Vaughn Jennings, Shelley Kath, Andy Kershaw, Jessica Kowal, Philippe Lalande, Priscille Leblanc, Susanne Manovill, Terry Marsden, Désirée McGraw, David McGuinty, George McRobie, Kevin Morgan, Karen Ocana, Denis Ouimet, Mark Rumizen, Doris Schröcker, Lori Stahlbrand, Barbara Turley-McIntyre, Wayne Roberts (who, among other things, suggested the title), Peter Schiefke, Frank Valeriote, Tom van der Meulen.

A nod to all my St Peter's Harbour friends who never tired of hearing about 'the book'.

For personal commitment, encouragement, help, and patience throughout the book's gestation: my sons Jonathan and Christopher, my sister Faye Stevenson (I did remember: *Mihi cura futuri*), Robert McIntyre, and all of my in-laws and outlaws.

My special thanks:

In all the ways that one would hope that a spouse would offer support, my wife, Dr Alison Blay-Palmer, excelled. But for her, that is just a starting point. Nothing suffices to describe the actual level of inspiration, assistance, and support that I enjoyed at her hand.

Abbreviations and Glossary

A4A	Airlines For America (formerly the Air Transport Association of America)
ABPPM	Associação Brasileira dos produtores de pinhão manso (Brazilian Jatropha Producers Association)
Absolute zero	The theoretically lowest possible temperature, 0° on the Kelvin scale, or approximately –273° Celsius or –460° Fahrenheit
ACARE	Advisory Council for Aeronautics Research in Europe/Advisory Council for Aeronautics Research and Innovation in Europe
ACI	Airports Council International
AEF	Aviation Environment Federation
AFQRJOS	Aviation Fuel Quality Requirements for Jointly Operated Systems
AIA	Aerospace Industries Association of America
AIC	Aviation-induced cirrus (or sometimes Aviation-induced cloudiness)
Albedo	A measure of solar radiation that is reflected rather than absorbed by a surface
ANA	All Nippon Airlines (Japan)
Annex 1 countries	Countries identified in Annex 1 to the Kyoto Protocol; these are the 'developed world' countries charged with primary responsibility for reducing GHG emissions
Anthropogenic	Caused by human activity
ARPA-E	Advanced Research Projects Agency—Energy (United States Department of Energy)
ASTM	ASTM International (formerly American Society for Testing and Materials) US-based international standards organization
ATAG	Air Transport Action Group
ATM	Air Traffic Management
Blend mandate	An imposed requirement for a specified minimum proportion of biofuel in any biofuel/petroleum-derived fuel mixture
BRIC countries	Brazil, Russia, India and China
BTL	Biomass-to-liquid
C	Carbon
CAAFI	Commercial Aviation Alternative Fuels Initiative

CAPA	CAPA Centre for Aviation (formerly Centre for Asia Pacific Aviation CAPA), Australia-based publisher of market analyses and data for the airline industry
CBDR	Common but differentiated responsibilities
CBSP	Council on Sustainable Biomass Production
CCS	Carbon capture and storage (or sequestration)—technologies that prevent carbon from being released into the environment as carbon dioxide pollutant
CH_4	Methane
CIA	Central Intelligence Agency (US)
CNG	Carbon-neutral growth
CO	Carbon monoxide
CO_2	Carbon dioxide
CO_2e	Carbon dioxide equivalent
Contrail	Condensation trail, visible water vapor left in the atmosphere by jet engines
COP	Conference of the Parties [UNFCCC participants]
CSBP	Council on Sustainable Biomass Production (US)
CSR	Corporate social responsibility
CTL	Coal-to-liquid
DARPA	Defense Advanced Research Projects Agency
DEF-STANs	Standards produced by DSTAN, the UK Ministry of Defence body responsible for standards, policies, etc.
DLR	Deutsches Zentrum für Luft- und Raumfahrt (German Aerospace Center)
DOD	Department of Defense (US)
DOE	Department of Energy (US)
EAG	Environment Advisory Group
EBAA	European Business Aviation Association
EISA	Energy Independence and Security Act (US)
EM	Ecological modernization
EPA	Environmental Protection Agency (US)
EPAct	Energy Policy Act (US)
EPFL	Ecole polytechnique fédérale de Lausanne
EU	European Union
EU ETS	European Union Emissions Trading System
EU RED	European Union Renewable Energy Directive
F-T	Fischer–Tropsch (a process that breaks down suitable raw materials into carbon, hydrogen, oxygen and other elements, then recombines them into liquid hydrocarbons that can be developed into synthetic fuels)
FAA	Federal Aviation Administration (US)
FIT	Feed-in tariff
FPIC	Free, prior, and informed consent

GAO	Government Accountability Office (US)
GDP	Gross domestic product
GDP-PPP	Gross domestic product on purchasing power parity basis
GE	Genetic engineering
GHG	Greenhouse gas
GMO	Genetically modified organism(s)
GTL	Gas-to-liquid
H_2	Hydrogen molecule
H_2O	Water
HRJ	Hydro-treated renewable jet
IASTA	International Air Transit Services Agreement
IATA	International Air Transport Association
ICAO	International Civil Aviation Organization
ICCAIA	International Coordinating Council of Aerospace Industries Associations
IEA	International Energy Agency
ILUC	Indirect land use change
IPAT	Impact, population, affluence, technology
IPCC	Intergovernmental Panel on Climate Change (UN)
IRI	Imperium Renewables Inc. (US company)
ISEAL	International Social and Environmental Accreditation and Labeling Alliance
JAL	Japan Airlines
KLM	Koninklijke Luchtvaart Maatschappij (Royal Dutch Airlines)
L/D	Lift/drag
LCA	Life-cycle analysis or life-cycle assessment
LCC	Low-cost carrier
LH_2	Liquid hydrogen
LUC	(Direct) land use change
MBM	Market-based measure
mHa	Million hectares
mio	Millions
MRV	Monitoring, reporting, and verification
MSW	Municipal solid waste
N	Nitrogen
n.d.	no date; publication date unknown
n.p.	no pages; pagination unknown
NASA	National Aeronautics and Space Administration (US)
NextGen	Next Generation Aircraft Management System (an initiative of the US FAA Federal Aviation Administration)
NGO	Non-governmental organization
NO	Nitric oxide
NO_2	Nitrogen dioxide
NOx	Collective term for mono-nitrogen oxides, NO or NO_2

NRDC	Natural Resources Defense Council (US)
O_2	Oxygen
O_3	Ozone
OECD	Organisation for Economic Co-operation and Development
Offtake agreement	An agreement to buy or sell a quantity of product, often negotiated before the product is actually available
PPM	Parts per million
R&D	Research and development
RF	Radiative forcing—the difference between the amount of energy received by the earth and the amount radiated back into space
RFS	Renewable Fuel Standard (US)
RPK	Revenue passenger kilometer
RSB	Roundtable on Sustainable Biomaterials
S	Sulfur
s.l.	sine loco; place of publication unknown
s.n.	sine nomine; name of publisher unknown
SAFUG	Sustainable Aviation Fuel Users Group
SARPs	Standards and recommended practices
SCC	Social cost of carbon
SCRC	Special circumstances and respective capabilities
SESAR	Single European Skies ATM Research
SRES	Special Report on Emissions Scenarios
Synth-bio	Synthetic biology
TPES	Total (global) primary energy supply
UCO	Used cooking oil (as fuel)
UK	United Kingdom
UN	United Nations
UNCED	United Nations Conference on Environment and Development
UNEP	United Nations Environmental Programme
UNFCCC	United Nations Framework Convention on Climate Change
US, USA	United States of America
USD	United States dollars
VAA	Virgin Atlantic Airways (UK company)
WCED	World Commission on Environment and Development
WMO	World Meteorological Organization (UN)

Introduction

This book deals with one aspect of a frightening situation. It talks about having a problem. But it is not the problem that is really scary; even children understand that problems often have solutions. It is rather that odd feeling that a child gets when it is clear that no one is in charge; that no one seems prepared to become seized of an issue, deploy the resources, and make the decisions. That kind of situation is unnerving indeed. As adults we know that we must either take charge of problems ourselves or do what we can to bring others together and start sorting out who can do what. There is some fear that the 'adults' are missing from the tableau. Our title, *Will Sustainability Fly?* is intended to admit of an open question rather than introduce its closure. But it seemed better than 'Where Are the Adults?' There certainly is an element of frustration in addressing such a question and writing a book like this, though. We do not have a book on how to solve the problem that I will describe. Rather, this book presents a deliberation on how to go about writing that other one. Here, the intention is to talk to those who have anything to do with commercial flight. The long list includes those who work in the industry, serve it, study it, or—and this is the part that brings in almost everyone else—those who use it, or would like to. What can be read here should interest anyone, and it is intended to be one small part of that exercise of asking people to think about who can do what. It will be about the way the sustainability challenge that we all face in other spheres makes itself particularly apparent in aviation. Because air travel is important to us, almost no matter who we are or what we do. And, as it turns out, this challenge of bringing sustainability to aviation, while evidently daunting, is probably the best shot that we all have in making a big, important, infrastructural, energy-intensive activity into something that stops harming our planet and its societies. This struggle could help us all learn a lot about the larger job. If we care, much of what is in these pages is very good news. But just like winning a prize, it is only good news if we follow and cash the cheque. We will examine things from the perspective of taking what we have and actually *using* it.

In a cautionary vein, though, establishing a balance between our capacities for biting and chewing is also important. Understanding that the problem about which we speak is huge, what is written here is limited in scope to one particular part of it: commercial aviation's quest for sustainable flight energy—fuel. The bigger problem can be described as that of sorting out how human activity can continue in the world without wrecking it. Our smaller bit (in that context) is: 'What do we put in airplanes to keep them flying? How would that happen?'

I find the topic thrilling. It is interesting on its own narrow merits but, even more interesting, it reveals so much about the bigger one. 'Will sustainability fly?' is not merely a word trick to reveal that we need to think about whether flying can be done in a non-destructive way. Rather, the resulting inquiry also uncovers so much about the more important question. In that respect, 'Will sustainability fly?' can be read as asking if 'Sustainability' is something that makes sense—in flying or anything else. We cannot know about flying's part unless we know what it is supposed to be part *of*. In broad strokes, we need to know how the topics of sustainability and fuel link to one another.

Fuel is an interesting thing. One way that we distinguish fuels is on the basis of where and how we get them: Do we pick them up, burn them, and then wait for more of a similar kind to grow? Or do we dig them from a fixed supply beneath the ground? Until a few decades ago we hardly made that distinction at all. Fuel was fuel and we only really needed to think about effort and expense in its procurement. Now we know that this distinction is critical. There are other ideas too. Fuel is not energy, it is a substance that *contains* energy—in most cases, chemical energy. One way of viewing the puzzle is figuring out if there is a way to capture different kinds of available energy and put it into the manufacture of a substance that we can use for fuel. So we are really talking about a sustainable source of energy for fuel. In other areas of human activity we can exploit energy directly. But while airplanes may come with batteries in the future, it will not be soon. Airplanes need portable, liquid fuel.

There are technical solutions to this largest component of commercial aviation's sustainability gap, but those solutions sometimes languish unnecessarily in laboratories and in portfolios of business cases, gathering dust. Here we will address our way forward in terms of how to get those technical solutions in place. Right now, we do not know that, nor what we should do to help make it happen.

Socrates gave us what we all now consider a common insight about the nature of questions: 'Understanding a question is half an answer.' What is presented here is information *about* a challenge. It is true that there are ideas and even suggestions sometimes too. But even these are proffered not as answers, but as ways of stimulating deeper thought about where we are and what we lack. What I would like to do is to help all of those who care to better understand what we are up against. This is a travel guide; our journey here is to begin to appreciate the nature of the job, its size, the reasons that it exists, the things that offer themselves as capable of addressing it, and what we have to do to make the necessary actions possible.

Why is fuel the thing? There is a lot more to aviation than burning fuel. True. But jet fuel is the single largest material input for powered flight. Each commercial airplane—whether relatively small and costing tens of millions of dollars, or large, and costing hundreds of millions—will burn through its own purchase price, many times, in fuel.

Thinking and motive in considering alternative sources of flight energy has been an evolution. There was a time, of course, when sustainability was not part of any equation that we acknowledged. In the past, the focus of alternative fuel was principally directed toward the problems of price stability and security of fuel supply. These remain key issues. But the industry seems to accept that new sources of fuel must perform better on criteria of sustainability. So, in the end, this book was written for reasons that are somewhat different from those that led to its conception. It has migrated some distance from being a simpler book about why and how an industry must change the source of its most important energy supply to a broader look at the nature and important implications of that change.

Several years ago, I was closing in on the end of a flying career that had me piloting commercial aircraft all over the world. But I was interested in environmental matters and stunned by the size and complexity of the problem that seemed to be arising around the matter of global warming. I was learning more about climate, so I knew that if the stories were true, aviation must constitute an enormous part of the greenhouse gas (GHG) emissions profile of people who spend time aboard airplanes, and that would include ('This is your captain speaking') me. Where does all this fuel come from, and exactly what is the effect of burning it? Everything about the business has effects: building the aircraft, powering the infrastructure. But my attention was captured by the fuel question. I became very interested in what (if anything) my industry was doing to address it.

I acknowledge that my initial interest was too narrowly environmental and, even more narrowly, about GHG emissions from the act of flying. Lots of people who end up interested in the larger ramifications start out thinking that way: about the environment. But when we think about a better world and a more benign presence for humanity in it, the discussion is now usually framed in terms much broader than environmental. We now routinely resort to that notable word 'sustainability'. Where did the matter of jet fuel lie in that regard? And, sure, maybe a different kind of fuel might allow aviation to have a lower impact on the environment, and in that way such a fuel would help flying to be more sustainable. But what of the implications (environmental and other) of *producing* such fuel itself? And how should our contemplation of all of these things be framed?

That gave rise to another more fundamental question: what exactly do we mean by 'sustainability'? After a lot of reading, I discovered—as others have as well—that our understanding and agreement upon its nature and meaning is not entirely settled. The term is employed in so many ways that it has been rendered mere noise in many people's hearing. The book is about a number of things, but the most important part of introducing the larger theme is to be explicit about how a discussion of the nature and understanding of sustainability in an industry context constitutes the root and core of everything that we must consider and accomplish.

It is an odd word when we use it in the current popular application that concerns the viability of the human project. There are certainly no easy substitutions for it. Still, in other respects, it is ordinary. The *Oxford English Dictionary* features meanings that relate to whether something can be perpetuated, supported and endured. Then there are ways of looking at those three characteristics not only in the physical sense but also in terms of justice and fairness. So we know what the word means: the same thing it has meant for centuries. It can be argued that when people ask what 'sustainability' *means* in current application, what they often really seem to be asking about is not definition but rather implication. If we ask,

'What does it mean to fail this examination?' we are asking about consequences. 'Failure' certainly means that we got less than the minimum passing grade, but what we actually want to know is 'What happens now, as a result of this failure?' So asking, 'What does sustainability mean?' could be, 'What happens if we do what we must to make the human project and its component bits supportable, endurable, and perpetuable?' or 'What has to happen to arrive at that desired result?' We are asking about conditions and consequences. It is time that we realized that the answers to those questions are simple in essence and complicated in application. The number of considerations is very large, and they affect one another.

And an odd metamorphosis has occurred as a result of our unrelenting use of this old word, sustainability, in the current global sense: We have so continually had reason to take recourse to it that the fact of its dual reference—to both physical and moral realms—has meant that we have not only started to weave them together but to understand (finally) that they *must* both be part of one fabric. So those of us who entered the subject with only the environment in mind have had to learn that the justice component must come with it or the environment cannot be saved—and why that is. As a planetary society, we are learning that perpetuating, supporting, and enduring in physical terms and terms of justice are all part of one, integrated whole.

We have all heard a popular formulation of sustainability as environment, society, and economics. That can seem arbitrary. Why these three things? Why not more? Why not fewer? But we are arriving at the relevant understanding of those questions. Society is us, and what we mean to ourselves and to each other; the environment is our physical milieu and our material support, creating, together, a very intricate dynamic. Economics is the way that we communicate value in that dynamic. If we do not understand each of the three factors in each of the others' terms, we are distorting our perception. That has a bearing on how we act within the dynamic, and it can create problems.

But an effort to make sustainability relevant and real in a particular part of the world's environmental, social, and economic milieu is not practical unless we understand the term in a consistent way. The only way that our understanding can be consistent is if sustainability is seen as comprehensively as possible; we cannot leave anything out. If we *do*, someone will disagree, and so they should.

That is a complex way of looking at something that some of us thought was simpler. Does anyone accept such a complicated approach? Thankfully, there are a great number of people and organizations that are well embarked and under way in making sustainability real and concrete for aviation. We must parse it, study it, examine it, and find the ways of putting it all together. That is what constitutes the efforts of those who are making the largest and most comprehensive understanding of the term as a real and assessable quality.

We preoccupy ourselves with definitions. When Einstein gave us *Relativity, the Special and the General Theory* (Einstein 1920), he didn't create 'relativity' as a new word with a new definition. And 'relativity' still means what it has always meant. Anyway it wasn't really a theory of 'relativity' itself, it was a theory about certain things that he hypothesized were actually relative to each other. Perhaps it should have been advanced as the 'theory of the relativity of mass, energy, time, and space.' I do not know. But in any case, he was just using an existing word whose meaning coincided with his ideas. Likewise, we do not need a new definition of sustainability; we already have one (many). We need a theory of how certain things can be understood in terms of whether they can be sustained. We know what 'sustain' means. When we say 'relativity' now, the context often lets us know that we are talking about Einstein's concepts. Likewise, 'sustainability' has become a bit of shorthand for a larger subject: Is human existence sustainable? What about component activities? How? In what form? Whether and how human existence and action can continue—whether they are sustainable, in the plain meaning— is still exactly what we want to figure out. I did not encounter any Einsteins in reading about and discussing sustainability. But I think that we should recognize the need to see how (what we are calling) sustainability might work and not just argue about what it means.

It is fair to ask why an aviation book seems to preoccupy itself with sustainability. Many readers may expect aviation itself to be the most important topic. The only answer that makes sense is that, regardless of anyone's most prominent *interest* (aviation, manufacturing, agriculture, or anything else), sustainability is the larger *subject*. Anyone picking up a book entitled *Can I Be Healthy?* will be principally concerned with the 'I'. But such a book will talk about what health is, and how to achieve and maintain it; 'I' will learn about health and understand what to do to adapt my actions to the pursuit and incorporation of health. That is what we are doing here.

But in some ways, the whole discussion has become much more complicated than it needs to be. When so many people think about so many different aspects of what we face in ways that are not necessarily consistent at all, it can get confusing. It started to seem to me that bringing the aviation fuel question to the sustainability context was like trying to jump on a train that was not only in motion but also still under construction; just not possible. I began to think that it was necessary to stop the train and try and get a version of it built in a way that would make sense to me, the reader, the aviation industry, and anyone else who cared to read. So now the book is about sustainability and whether aviation's flight energy needs can be met in a way that conforms with our discoveries about its exigencies. The effort has been to change our understanding of sustainability from a shape-shifting chimera to an idea that should make sense to everyone who currently uses the word in different ways, and then to take on the question of whether we can bring this complex quality to the process of securing flight energy.

Whether one is a pilot or passenger, it gets personal. The experience of flying an Airbus 340-500 from Toronto and over the North Pole to Hong Kong, for example, involved the routine consumption of well over a hundred *tonnes*[1] of jet fuel—

---

1 A tonne (metric ton) is 1000 kilograms, or 2200 pounds. A North American ton (short ton) is 2000 pounds.

much more than would fill the average backyard swimming pool. Where does it all come from? Currently, jet fuel is sourced almost entirely from petroleum oil. (I say 'almost' because the first tentative steps to bring lower-carbon fuels into flight use have already begun.) Even in the larger scheme of things, that is a big deal, and it cannot continue. That is our particular job.

But aviation alternative fuels is a revelatory example of the complex challenges faced by the larger world, and it is my hope that we can begin to understand that a simplistic approach to the question of sustainability will produce a weak result or even make the situation worse. While acknowledging this caution, the main message here is that we *can* make things better—and we need to. Perhaps the way that aviation understands and pursues the goal of sustainability will help others.

One thing that makes aviation a wonderful entry point to the larger discussion is the relative absence of issues of national vested interest. While national interest always plays a prominent role, the question of aviation's sustainability is slightly less wrapped up in the matter of advance or decline of specific countries than other industries tend to be. All countries use air travel, and it supports *all* economic activity. Air travel is important to everyone, and while it may be more important to some countries than others, we do not see that to the same extent as with many other parts of world commerce. And while we will see that air travel is a relatively small contributor to atmospheric carbon dioxide (CO_2), it is a big user of fossil fuel on a unit basis, so it is a great place to try to make headway.

Another reason to reflect on aviation's status as a prime candidate for emissions-reduction initiatives is the attention that it draws. Air travel has a remarkably high profile. It elicits comment out of proportion to its environmental impact. While most CO_2 production in the developed world comes from burning fuels to do other things (heating, cooling, and power to run our buildings, machines and vehicles), producing cement (buildings and roads), and producing food (agricultural machinery, fertilizers, deforestation) (Le Quéré et al. 2013), seeing that one large aircraft overhead on its way to the other side of the planet seems a particularly profligate bit of environmental abuse. As a consequence, there will be mounting pressure to impose sanctions against air travel emissions. The industry will hope that those policies come with support for the sector's attempts to improve. Otherwise, costly measures to penalize emissions may eat up the cash required for the development of technologies that could achieve the goal of reducing them. Aviation has been making a long and expensive effort to improve fuel efficiency (Penner et al. 1999, Intergovernmental Panel on Climate Change 2014) and although this effort continues, the sector will certainly want help in making the further order-of-magnitude improvements that will now be required.

Some of the important actions in this regard may have less to do with *how* air transport is accomplished and more to do with *how much* of it is. We have known for a long time now that reducing global warming will be a long, complex task—perhaps involving transportation mode shifts and a number of other things. Before anything else, we must acknowledge the need to grab the low-hanging fruit. Current strategies and policies are explored in Chapter 8, Transport, of

Climate Change 2014: Mitigation of Climate Change (Intergovernmental Panel on Climate Change 2014) and they include shifts to other modes of transportation and communication. However, we still need airplanes. How can we make them less damaging to the environment in terms of GHG emissions but also in other respects? Confronting the reality that aviation is not disappearing at any time soon, we start to wrestle with the question of how its impact might be lessened. Efficiency? Yes—absolutely. But it is hard to make flying more efficient—some of our most talented engineers have been working on just that for over a hundred years. While there is some substantial room for advancement in aircraft and engine design and in airspace management, nothing offers the revolutionary improvement in emissions reduction that environmental considerations must ultimately demand. A new kind of airplane? That will probably come. But the start of a complete alteration of aviation's energy regime is at least a generation away. A new kind of aircraft might be fueled, just for example, by liquid hydrogen produced using, say, renewable sources of electricity. But such changes are not envisaged at all in the transport trajectories to 2050 assumed by the International Energy Agency (International Energy Agency 2012). And even if we could introduce a new sort of aircraft technology tomorrow, it could not be built up to numbers sufficient to replace the current fleet for many years. So aircraft like the ones we have now will be around for a long, long time; what can we feed them?

We have to figure out new ways of getting the *same* fuel for them, but without continuing our dependence on fossil petroleum. This is the part of the discussion that is so encouraging. Science presents a few avenues; a prominent one involves converting biomass to jet fuel. Admittedly, many people will be startled to think that the aviation industry would consider trying to survive in a non-carbon world by resorting to things like biofuels. We are all aware of some biofuel experiences (notably corn ethanol) that have resulted in a certain amount of failure and disappointment. The fact that using corn ethanol as a substitute for gasoline so marginally reduces GHG emissions (if at all) has left the impression that biofuels cannot help. Clearly, if biofuel is part of the solution, it cannot be the kind that we have already seen, and we cannot make assessments about its sustainability the same way that we have in the past either. That is true. But we will see that this fact is also well recognized by those in the aviation and fuel industries, and that recognition is itself encouraging.

Then again, anyone familiar with the air industry might wonder how biofuels or anything else can be substituted for a specific kerosene hydrocarbon blend that airplanes, each constituting an investment of perhaps hundreds of millions of dollars, have been carefully designed to burn exclusively. To answer that question, techniques have been developed to turn various types of biomass into 'paraffinic kerosene' indistinguishable from the main component of conventional jet fuel. This is important because it *is* true (in practical terms) that commercial jet aircraft can only burn current types of jet fuel. Replacements do not take the form of a 'substitute'—say, vegetable oil—but rather a 'drop-in' product that is *just like* regular old jet fuel. It might start off as vegetable oil, but means have been developed to turn it into the appropriate hydrocarbon. Blends of up to 50 percent synthetic biojet fuel are currently in use on a very limited basis (Lane 2012). So a lot of the technology is now increasingly well understood—although it is constantly evolving, broadening, and improving. In fact, some of the most interesting research is into sustainable alternative fuels that are not necessarily biofuels at all. The current challenge is to scale up production, processing, and distribution infrastructure, and to fund the development of even better fuel technologies.

But the question about the very viability of the whole concept of renewable fuels is more interesting: can any of this be accomplished in a way that is really environmentally useful? Fortunately the answer seems to be, Yes—if we are careful. And the air industry seems well aware of the care that will have to be taken because there will be a great deal of scorn if alternative aviation fuels provide only marginal advantage in reducing GHG emissions, or if they bring about other new types of environmental degradation.

So, perhaps, it *can* get done. Right away? No. And that fact is at the center of the debate. Not the technology debate, the policy debate. No matter how much more quickly we would like to change, there are also good reasons why it will not be done without an incremental approach. Some of the best potential fuel technologies are the furthest away from deployment, and in trying to decide on alternative fuel options, major players must address many factors having to do with that relative readiness. The best solutions to the problem will not necessarily be the first ones deployed. And again—very importantly—we shouldn't think that GHG emission is the only consideration here: the air industry has embarked on this pursuit of sustainable fuel with the understanding that invocation of the word 'sustainable' brings a more comprehensive onus than was borne by the supporters of renewable biofuels back in the 1980s.

All of this will be hard work. But it cannot be allowed to daunt us. We have to understand that the initial moves in the direction of sustainable fuels will be tentative, and will not satisfy the ultimate emissions reductions and sustainability goals that the industry must set. Sustainable aviation will be slow to get off the ground (smile). So a first realization is that there must be some room for evolution of understanding and effect. But getting started and speeding up are critical.

There will be challenges. The 'food for fuel' question is a huge part of the sustainability discussions and will not disappear. Many people will argue quite reasonably that using land for fuel production is never all right if people are starving. Others will say that as long as the land and other resources are not being used to produce food right now anyway, it does not really matter. How will our policies and our ways of assessing sustainability react?

And speaking of policy, another thing that the book explores is the historical policy context within which aviation commerce, on the one hand, and global GHG emissions, on the other, have been discussed. Policy on sustainable fuel may be a fairly new area of interest, but we do have other policies relating to how commercial aviation works and how emissions should be regarded. We will see how important principles such as the equal and non-discriminatory treatment of air carriers as a consequence of discussions at the Chicago Convention in 1944 compete with the United Nations Framework Convention on Climate Change (UNFCCC) principles of Common But Differentiated Responsibilities (CBDR) for states, described 50 years later. Tricky stuff.

Much of the discussion about aviation and emissions at the international level now is about reducing emissions through charges or, alternatively, accounting for them through offsets. Obviously, in the best outcome, carbon charges and restrictions become irrelevant. But carbon charges, or market-based measures (MBMs), may play a critical interim role in achieving our goals. If they are implemented, can that be done in a way that gains universal support?

And if we accept that international policy discussions can be complicated, who else can move things along? What about the airlines themselves? Can they be relied upon to undertake action unilaterally? Well, yes—to a degree. The brightest stars do seem to recognize the long-term importance of getting away from the fossil fuel track. A strong international institutional commitment to sustainable biofuels is now apparent in the industry, with air carrier representation groups like the International Air Transport Association (IATA) showing increasing policy support for and commitment to sustainable fuel (International Air Transport Association n.d.). But the particular degree of commitment that individual airlines are likely to demonstrate in securing supplies of more sustainable jet fuel depends on a few things.

The first is the strength of its balance sheet. While cost of fuel is an extremely important factor, currently about 30 percent of costs industry-wide (International Air Transport Association 2014), long-term strategy and a cash dedication to securing alternative fuel at some point in the future is not necessarily appreciated by some boards of directors and is not even in accord with the time frame which most carriers do contemplate when, for example, *hedging* fuel.

The second factor is an airline's managers' knowledge; if there *is* thinking ahead about alternative fuel sources, it is this group's understanding of which resource and fuel processing technology might best serve its interests, and (again) in what time frame, that counts.

The third has to do with the 'vision thing': an airline's leaders will reflect some particular blending of national, corporate, and individual level of imagination and creativity. Some Chief Executive Officers do get it. But the complete re-conceptualization of the industry's fuel supply is, like the response to deregulation in the 1980s, an exercise in collective mind opening that most technocrats will find daunting, despite the industry association consensus on the need to move toward sustainable fuel. Concerns include environmental liabilities, security of supply, and stability of price as the most readily exploitable deposits of petroleum are played out amidst rising demand. So interest is high, but the individual actors in the market (potential fuel suppliers and individual airlines) are having a hard time moving the fuel agenda by themselves.

In the context of the foregoing, it is not a surprise that air carrier interest in alternative fuels has focused not only on the cheapest but also the most readily available options. But airlines' focus on immediate benefit puts longest-term sustainability issues at risk. Policy must take a longer view. We need to assess, in a comprehensive way, not only the relative environmental but also the social aspects of the production of fuel. It is understandable (and even desirable) that individual companies would focus on something that offers an immediate, albeit limited, environmental benefit, but someone has to be looking farther down the road to see where we should be headed in the longer term and taking concrete action to make that happen. It is becoming routine to see press releases about an airline's commitment to the alternative sources of fuel. Policy development bodies must recognize that there are key players in industry who recognize and accept that short-term solutions are of limited value but who (like the rest of us) lack that crystal ball.

Alternative fuels from renewable sources *are* being developed and some are starting to be commercialized. Some of these initiatives represent strides in progress toward levels of sustainability. This makes some people think that there really is no need to do anything. There is. The problem is pace of commercial development: insufficient numbers of projects reaching commercial viability at too slow a speed. What policy related to development of sustainable aviation fuels should look like is very much an open question at present, but some things are fairly clear, and this book will make a case in the following terms.

The first is that the ideas that stand to deliver fuel in the most sustainable manner at the lowest cost are currently just at the research horizon. While such ideas do attract investment capital, they do not enjoy nearly enough support to move at an appropriate speed into deployment and displacement of poorer options, while those options themselves climb the difficult hill of building commercial, infrastructural, economic, and political commitment. To address this we need significant support for research, development and deployment of new fuel energy technology.

Second is the need to provide commercial incentive. But if policy must encourage success, it must also be prepared to sanction failure. Commercial entities can be limited in perception; they see market value almost to the exclusion of other factors. The externalized costs of carbon emissions are certain to be ignored until they constitute an imposed negative commercial value. There is little doubt that we absolutely need to develop more comprehensive and linked carbon pricing mechanisms, and we need to do it expeditiously; failure to help our world economy see the need to move into a post-carbon phase is very much a detriment.

Third, since key players, such as airlines, are sometimes blinkered by inherent short-term focus on current cash flow, only the largest of our institutions are in a position to assess the infrastructural value to the economy of major (and initially expensive) technology shifts. Only government can have the economic and policy-generating weight to help develop these brand new technologies and then move them into the commercial sphere fast enough to create a relative advantage.

Anything worthwhile that we do must result from frank and honest assessment of the demands of our economy. The marketplace will certainly play its role in bringing more wealth and prosperity to the world *as long as* the market has all of the information in terms of the real cost of doing the wrong thing.

Certain governments *are* reacting, and this discussion gets beyond one country and one mode of transportation; around the world there is burgeoning interest in biofuels for all sorts of different applications. Heavy, long-haul road transport will not be able to get away from liquid fuels any time soon; there is just a lot more energy per unit mass in liquid fuel than in batteries and hydrogen for fuel cells at present. However, the air industries are especially trapped by their absolute reliance not only on liquid fuel, but also on such a very *particular* liquid fuel. For that reason, some governments are starting to react. In the United States (US), where the military establishment is the largest single consumer of energy, it was a Defense Advanced Research Projects Agency (DARPA) grant that helped UOP, a Honeywell subsidiary, to develop biomass-processing technology that made transmutation of the mundane byproducts of life into high-quality hydrocarbon fuels, possible—something that geological processes had taken millions of years to accomplish in producing petroleum (Warwick 2007). That initiative took place several years ago, and the US military is now fostering commercialization through procurement initiatives. Other research spending on alternative fuels technology is continuing to grow in the developed world. Even in the less-developed world, there is substantial support for alternative fuels development. China is committed to progress in developing advanced biofuels (though there are some doubts that it will be able to reach its ambitious goals 2011–2015 due to insufficient supplies of feedstocks) (Riedel, Scott, and Junyang 2013). And while massive levels of spending in superpower economies is almost inevitable in *any* area, one only has to examine the level of sophistication in analysis going on elsewhere to see that other countries are unwilling to be left in the dust. The United Kingdom's Commission on Climate Change is a good example; it publishes an annual summary of the air sector's GHG emissions and energy needs (Committee on Climate Change 2013) and many other useful reports.

The biggest challenge will be to elaborate the meaning of sustainability in the context of alternative aviation fuel and to have a broad international commitment to understanding, measuring, and demanding comprehensive subscription to various elements of sustainability. This will be a difficult challenge, since each facet of the broader concept affects each country in a different way and they will fight for advantage. But if we are not able to come to agreement, the air sector will be burdened with the costs of expensive offsets or taxes.

To outline what you are about to read in a more ordered way, we start with a discussion of the extent of aviation GHG emissions. It is necessary to see what contribution the sector makes to the global GHG problem, critically, to gain a sense of the way that GHG emission is distributed among individuals. Then we examine the likely evolution of this picture.

Confronting the challenge, the book takes on the questions about how we can make progress in reducing aviation GHG. Efficiencies to be incorporated in aircraft design, operational practice, and airspace management are seen to be crucial, but not enough. We examine the factors that affect development of completely new aircraft technology and why we are (still and nevertheless) left with the need to continue with aircraft that are a lot like the ones that we have now.

We then talk about why current aircraft and the newer, improved designs of similar ones cannot be asked to burn something other than kerosene (paraffin). This all eventually yields to a key question: Is there any other way to produce kerosene jet fuel? It is possible; it is being done. We will explore many of the developments that are occurring in new fuel technology so that the reader has some sense of what is possible.

All of that brings us to the threshold of a deeper discussion. In order to set the scene, we provide background on an essential element of this project: policy. What policy have we developed over time that might have a bearing on our discussion about what should happen now? Established policy is not about new fuels. It is about how international aviation is commercially governed, on the one hand, and the limitation of GHG, on the other. At some point, those two things must operate together in a way that fosters solutions to the challenges that we face.

If we identify 'sustainable, alternative sources of flight energy' as the most important tool that will *allow* us to face our challenge, the most critical thing is to understand the task in a coherent, rational, and consistent way. That is why the discussion about the word sustainability and its current popular usages is so important, and such a large part of this book and all of the discourse that will attend any of the subject matter in whatever forum. Our inquiry into how we might 'measure' the sustainability of fuel will reveal even more insights into the developing consensus on what it means and how it operates. This is an encouraging part of our story.

The last part of the book undertakes an examination of how those things that are physically, socially, and economically necessary can become possible; how understanding and need can be turned into policy that will bring the current evolution toward sustainable fuel to a pace and in a way that will serve the needs of the industry and society in all parts of the planet.

Finally, we talk a little more specifically about what is actually happening and meet some of the people, organizations, and companies that are part of the change. Throughout the book, I have depended upon comment and input from a number of people who are active observers of or participants in the effort to move us toward commercial inevitability for truly sustainable liquid energy for flight.

A 'Fasten Your Seatbelts' caution is not out of order: our first chapter will describe the size of our task.

Chapter 1
The Size of the Problem

Scope in Time

In order to get where we are going we sometimes have to go somewhere else first. Maybe we even have to start off in a different direction: the train goes east to our destination but the departure station is at the west end of town. The matter that is discussed in these pages is enormous in its complexity and consequence, and moreover is only one part of a larger difficulty that is almost beyond our ability to conceptualize. In those circumstances it is too easy to get focused upon what is of most immediate concern and, taking a step in what seems like the right direction for now, we find that we have left ourselves a longer and more difficult journey. So we need to take the long view in examining the question of how aviation will address global warming and climate change. Some of what follows will seem of little practical applicability to our situation in this very early part of the twenty-first century. But we cannot—we dare not—lose sight of the fact that we have to solve these problems in a way that will have made sense at the *end* of this part of our history. What is offered here is presented in the context of striving for an ultimate good in an ultimate time rather than pursuing some more immediately advantageous tactic.

This chapter is specifically about the scale of difficulty that we confront. In order to consider that, we should bring a few of the remarks contained in the Introduction into more direct context here.

We face an entirely new kind of problem in the issues that bedevil our world's climate. The deployment of a fossil energy resource into a technological and social revolution in what is now the developed world, over nearly two centuries, creates a difficulty. But the whole world clamors for development. Maintaining living standards in the developed world and enabling others to improve their lot now requires us to leap ahead and put that same technological prowess out in front of what we have created. While this is not an uncommon sort of human circumstance, it is unprecedented at a global scale.

Depending upon technological fixes when it is technological innovation that brought us to this pass is, admittedly, pretty unnerving. When we think about our climate wobbling, the pace at which technology is developing, and the different directions that it seems to want to take us, we are entitled to feel that we have a very uncertain hold on this tiger's tail and we are being pulled off our feet. But no matter how distasteful further recourse to technological fixes may seem to some, we have to put every last bit of energy and wit into allowing our technological strength to perform as it has never performed, to achieve as it has never achieved.

Events have foreclosed on many other options. It is not certain that we will succeed, but we will definitely fail unless we can move rapidly and with purpose to somehow get the tiger caged. A few of us may be injured in the endeavor.

By contrast, many are instinctively (and justifiably) tempted to simply throw up their hands and cry, 'Enough! Stop everything, development has become evil!' However, if it is true that we have been running heedlessly, it is still also true that if we stop or veer too abruptly, the danger of falling and landing on our heads is real and immediate. In any case, it is unlikely that the aviation industry, the people who now use it, and all of the people who would *like* to use it would contemplate just stopping. If we are willing to consider the premise that we cannot halt, and that we need to harness our economy and technologies to work our way out of the current threat, we should also expect that caution must now become a much more important consideration. We have to do something, but first we need to spend more effort to understand the problem in a detailed way. The need for care in determining our actions is as great as the urgency.

As to the reality of all of this, the overwhelmingly common assessment remains that the threatening climate crisis is real. Politicians seem to accept the current science-based assessments—notwithstanding a few frequently uttered, politically motivated, and publicly stated opinions to the contrary. Our leaders may wrangle and opine over who is most responsible and who among them is best equipped to take initiative and direct the use of money and resources, but they do not spend a lot of time denying that our world's atmosphere threatens to change the fabric of climate and life.

While we remain unsure of exactly what to do about global warming and the climate change that will accompany it, recognizing the need to acknowledge and react is the only useful way forward; that is becoming a political and social reality. I do not set out to prove the validity of current climate theory, I simply focus here on the place held by aviation—particularly commercial aviation—within the climate discussion. We want to know how aviation contributes to the problem and how it intends to reduce its negative effects.

Arguments presented here regarding technology, economics, and policy make a certain assumption about a larger perspective: Recognizing that the climate crisis represents failure to ensure that our actions could be perpetuated without compromising the status of our planet's ecological systems, we must now proceed on the basis of an informed commitment to the notion of sustainability. Nothing that we intend to do in science, commerce, or any part of the sphere of human affairs (including our developing struggle toward a remedy for our signal failure in the matter of climate) should be undertaken unless the proposed action is demonstrably sustainable in all of sustainability's dimensions, or at least fits within a strategy that leads *toward* sustainability. If it seems that addressing all of the provisions of a sustainable approach will slow us down, we have to remember that if our actions are *not* sustainable, we will end up worse off than we are now. We have ignored many of the negative externalities, the unintended consequences of what we do. It is time to account for them, and it is beyond time to ensure that when we are embarking upon any new path we ask these questions: What are we doing? Why are we doing this? What will happen?

As this book proceeds we will gradually knit as many of the elements of sustainability as possible into the fabric of our discussion. But it is air travel's special relationship to our climate issues that inspired this project, and it will be our entry point.

I qualify what follows with the foregoing because talking about the size of a challenge is not useful if we do not think about its nature as well.

The Raw Numbers and the Raw Reality

First things first: How big is aviation's contribution to the perturbation of our atmosphere? As later parts of the book will show, policy—and the inevitable relative urgencies that attend its discussion—demands a sense of scale in order to resolve matters such as the setting-out of national and international priorities over time. It is important to know the extent and nature of the contributions of air transport and other industries to our climate difficulties so that we can determine pace and priorities, and assess and allocate the resources that we might bring to bear in addressing them.

Without going too deeply into the science of global warming, paying some attention to the physical mechanisms will allow us to contextualize our thoughts on how we should react. Air travel does not contribute to global warming and climate change in exactly the same way as other industries or activities do, so it is important to understand its differences and the ways in which it must change. It is a complex task to assess the effect that commercial aviation has on the ocean of air that blankets the Earth and the life that inhabits it.

The aspect of this discussion with which we are probably most familiar is the production of greenhouse gas (GHG), most prominently carbon dioxide (CO_2). All GHGs have a residency time in the atmosphere, and that of CO_2 is quite long. A good part of the CO_2 that has been emitted since the dawn of the industrial age is still up there. When we look at the contribution made by an industry or by a country's overall economic activity, we must consider not only the current rate of contribution but also the persistent remnant of the total contribution to date. Aviation's current rate of contribution of CO_2 is not only fairly modest, but has only become significant in the last few decades, so the amount of CO_2 currently in the atmosphere that can be attributed to commercial flying is extremely small.

But CO_2 is not the only pollutant that is emitted in air travel. Water vapor, aerosols, oxides of nitrogen, oxides of sulfur—many substances are cast into the air by airplanes and other emitters. However, important differences are associated with the special nature of aviation: even though other human activities produce some of the same pollutants, airplanes deposit them at altitude. What is the effect of releasing these substances in the stratosphere, several kilometers above the Earth's surface?

We must set out here knowing that our answers can be framed in only limited detail. It is a challenge to accurately quantify the amount of CO_2 for which airplanes are accountable, but even more difficult to assess the amounts and altitude-dependent effects of the entire host of materials that come out of our collective air fleet's tailpipes. And let me acknowledge that advocacy can seem to alter the facts; it is not strange that environmentalist organizations can find numbers that differ from those offered by the air industry or its regulators. But most of the differences take the form of variations in what gets emphasized in news media rather than actual disagreement over the data. There are reliable sources that both industry and environmentalist advocates use in assessing and forecasting emissions and their effects.

The Significance of Data

Since much of what we talk about here is hard to quantify and controversial at the level of detail, it is important to consider what we really need to know. It is certainly necessary to have a clear picture, but we do not need every number to be rendered with great precision; we often only need to be able to say (albeit, with very high confidence) whether an effect is big or small and what its particular character might be. Since the technological and economic context that provides the setting for *all* future polluting activities is dynamic, we must accept a coarse assessment of aviation's contribution to global warming and climate change, partly because our understanding of so many aspects is inevitably tentative. Many factors are rapidly evolving: the science of the character of emissions, the technologies that we might employ in reducing them, and the future extent of air travel activity in absolute terms and (perhaps even more importantly) in terms relative to the size of the global population and the total economy. But we must know with certainty whether aviation's contribution is considered to be large or small, how it is likely to trend, and what is its nature.

Personal Shares of GHG Emissions: Flying is a Sin

In the summer of 2006, during the height of holiday travel from the United Kingdom (UK), the Bishop of London observed that a life characterized by profligate carbon emissions was not right, and that flying away in an airplane to enjoy a vacation was a particular sin (Barrow 2006). It would be understandable if the hearts of travelers and air travel workers dropped at that news: a single flight looms large in an individual's GHG emissions profile and the world is wrestling with a serious climate matter.

According to 2010 data, the average Canadian or American (to use high-emissions developed-world examples) produced on the order of 20 to 22 tonnes of all GHG emissions as deemed equivalent to CO_2 per year. But UK 2010 per capita emissions were much lower, about 10 tonnes (Conference Board of Canada 2013). The same *Daily Mail* article that outlined the good Bishop's chagrin also describes how a single flight can substantially raise a person's GHG emissions: a family of four traveling from the UK to Florida and home again would be responsible for an additional 6.4 tonnes of earth-warming pollution—1.6 tonnes for each individual (Barrow 2006). Assuming this vacationing British family's GHG emissions were in other respects average, that one additional flying vacation would raise its profile by approximately 16 percent. We can extrapolate the rather startling reality for those among us who travel frequently by air to distant places. It is probably part of the reason why the numbers for Americans and Canadians are so much higher.

There is no question that air travel can be a potentially very important part of an individual's personal carbon footprint if they fly, especially if their GHG emissions are otherwise modest. So it is no surprise that aviation causes a particular fright in those parts of the developed world where significant initiatives have been undertaken to reduce overall emissions. Similarly unsurprising, policy-makers in the less-developed world, where per-person emissions are quite small due to lower levels of economic activity and attendant lower standard of living, can see air travel as an especially offensive extravagance, a costly luxury for wealthy people in wealthy nations. And yet, many in the developing world want greater access to air services for themselves. While emissions from air travel are low globally because of the relatively small numbers of people who undertake it, they would be huge in a world where our development goals for less-wealthy people and countries had been realized.

Ironically, air travel is an important facilitator for the kind of economic development that is often fervently sought, especially by the least advantaged among us. The extent to which aviation functions in world and state economies in an infrastructural, enabling, and wealth-producing role, and the extent to which access to air travel constitutes a social and political justice issue will be explored in later chapters. But it is important to note here that airplanes serve a role in society and in the economy, beyond delivering cash to airlines' balance sheets.[1] The air industry makes claim to a disproportionately large contribution to economic activity, a claim that is sometimes overstated, according to critics, but generally valid. Nonetheless, the larger economic benefits of air travel should not be used as an excuse for inaction on the environmental front. If the general effort to reduce GHG emissions intensifies but aviation's share is not contained, then other sectors will have to make larger reductions in order to achieve a continuing decrease in total emissions. Will it help the economy if aviation gets a free emissions pass while every other sector suffers inordinate pressure?

Global Aviation Emissions

As we shift from looking at the degree to which a flight can affect an individual's emissions to the question of the size of aviation's contribution to the world's

---

1 Some airline executives would argue that air travel has *never* delivered cash to airlines' balance sheets.

global warming difficulty, it is encouraging that the airline industry has been able to make great strides in making flying more efficient and keeping its total contribution small. Nevertheless, as we develop the discussion, we will find two additional things: unless something is done, air travel will be an ever-growing part of the problem, both in absolute and relative terms; and that the state of current knowledge does not allow us to be entirely sure how big the problem actually is.

I have avoided depending upon any sole data set and this presentation should be read in that light; the sources quoted are those that I find to be credible and representative. This kind of qualification will appear more than once: It is probably unhelpful to argue over particular data when the point can often be more usefully understood in terms of a range of possibilities.

Aviation is a growing and different kind of polluter. By one estimate, aviation's current rate of emissions of CO_2 constituted 2–3 percent of the anthropogenic (caused by human activity) total for 2005 (Owen, Lee and Ling 2010). But the same study also points out that the raw CO_2 number is only part of the picture; other exhaust substances plus the consequence of these substances entering the atmosphere at altitude represent a potentially larger total global warming effect than that of CO_2 alone. The sum of the various ways that aviation pollutes can roughly double its global warming effect, rendering it on the order of 4.9 percent of humanity's current contribution. The other statistic of note in this study relates to the size of the aviation industry: recent annual growth in the sector has averaged around 5 percent. The research on the nature of the pollution that aviation contributes must be considered together with the forecast growth of the industry, to provide an overall picture—but it is a rather blurry one.

The global warming effect of a pollutant—when the character of the atmosphere is modified in a way that changes the balance of incoming and outgoing radiation—is generally referred to as radiative forcing (RF).[2] There are several RF effects. Each is difficult to quantify. Accordingly large variations exist within the bounds of current estimates for the total aviation RF.

As for industry growth, it is difficult to settle on a pattern for the future; while such forecasts exist—and we will discuss them—they do not necessarily delve into all of the factors that might conspire to confuse the outlook. There is also the fact that we cannot be certain about prospects for the larger economy and aviation's situation within it. In summary, while the actual quantity of aviation emissions is known to be small, its precise extent is not agreed upon; it is growing, but its future size is difficult to predict; and it may be more significant than its raw CO_2 number alone would suggest.

If that were not enough uncertainty for now, we also have to remember that since policy discussions will emphasize aviation's *relative* contribution, whatever the air travel industries might accomplish in reducing emissions must be set in a

---

2   For a comprehensive discussion of current assessments of global radiative forcing, see Myhre et al. (2013).

context of what the other growing human enterprises might accomplish in reducing the effects of emissions in other areas.

What we have mentioned so far puts a host of things in play. The rest of the chapter will first explain a little bit about the mechanisms of global warming that bear on aviation's emissions profile, then give some size and shape to the effects created by CO_2 and other substances. This discussion of how aviation contributes to global warming, together with comment on aviation's anticipated absolute growth, will give us an idea of how big an issue this gets to be in absolute terms if we do nothing. It also inevitably leads to a discussion about some of the things that the industry can do to reduce its impact in other ways. The chapter concludes with some data that deal with growth in global emissions from all sources so that we have a picture of aviation's possible relative contribution.

The Mechanisms of Global Warming in an Aviation Context

Allowing for some simplification, it is now necessary to describe global warming in more technical terms in order to provide an idea of how aviation's contribution can be measured and its relevance understood.

Obviously, the challenge of global warming is the management of the things that cause it. However, those who would act must first understand something about what is going on, and aviation's global warming profile is among the most complex. Any non-scientist layperson (including this writer) would find the subject difficult. For example, the huge effects created in the atmosphere by the addition of relatively small amounts of routine substances are hard to understand, at first. But they can be grasped sufficiently for our purposes here.

One thing that perhaps helps to make it clearer is a point concerning 'marginal effects'. It is important to recognize at the outset that all of the processes described here play at the margins of the enormous flows of energy into and away from the Earth's atmosphere. Even the threatened few degrees of global warming represent a critical change, so subtle effects are significant. The next paragraphs reflect some of what is described in current and authoritative documentation as regards the scientific basis for climate change (Le Treut et al. 2007).

In the simplest terms, if the average global temperature is in a steady state, the amount of solar radiation that falls upon the Earth must return to space. All energy that the sun sends to our planet in whatever form—as light or any other kind of radiation—is converted to heat when it strikes and is absorbed by the Earth, which then constantly radiates that heat back toward space. Some of it escapes immediately, and some is held in the atmosphere for a while; the Earth's average air temperature is a measure of the total amount of heat energy that resides in the atmosphere. If the heat energy that the Earth radiates back into space through and from the atmosphere is equal to the total energy that the sun sends here, the temperature of the atmosphere remains constant. But if there is any increase in the amount of heat energy that becomes trapped and absorbed, the atmosphere's temperature starts to rise. At some new higher temperature the system will again stabilize: the amount of energy coming in will equal the amount going out. Greenhouse gases are so called because their properties increase the atmosphere's capacity to trap and retain heat radiation. GHGs are essential; we would freeze without them. But too much is a problem.

There are several specific mechanisms that act together to produce or alter a heat balance, but those that concern aviation emissions fall into three main categories: the effects of GHGs, then of substances that *destroy* GHGs, and also of albedo. This last is the planet's reflectivity (a mechanism that returns solar energy back into space before it has a chance to be absorbed by the Earth and be re-radiated).

First, let us give some relevance to the previous point about marginal effects. If the whole discussion of global warming concerns the total quantity of heat in the atmosphere and whether it is changing, it is necessary to understand that many of the factors in the atmosphere's energy balance are not important in the context of global warming because they are stable or unchanging, and so do not alter the Earth's atmospheric temperature. Only the things that result in *change* are going to matter here, even if that change at first seems unimportant. For example, some people are mystified by the case for global warming based on atmospheric CO_2 when there is so little of it that it is measured in parts per million, and human activity raises that amount by only a fraction. However, standard atmospheric temperature at the earth's surface, 15° C, represents a total of 288 Celsius degrees above absolute zero (−273° C, the accepted definition of a complete absence of heat energy). Global warming is considered to be significant if it increases the atmosphere's temperature by only two Celsius degrees, which is less than 1 percent of the atmosphere's current total number of Celsius degrees above absolute zero.[3] So when we think of climate change, and the global warming that will bring it about, we have to remember that small changes in the amount of energy contained in the Earth's atmosphere are extremely important relative to the narrow range of temperatures that the planet has experienced since complex life evolved on the surface.

Realizing the importance of small factors, let us discuss the three categories of effects mentioned earlier. First, the heat-holding RF effects of greenhouse gases themselves: we have already touched on those. Second, there are substances that destroy GHGs, obviously causing a *reduction* in global warming effect; RF values of these substances are generally expressed as negative numbers. The third effect, albedo, is more complex in that it concerns total reflectivity. Energy reflected by the

---

3   For those more familiar with Fahrenheit: Standard atmospheric temperature at the surface, 59° F, represents a total of 519 Fahrenheit degrees above absolute zero (−460° F, the accepted definition of a complete absence of heat energy.) Global warming is considered to be significant if it increases the temperature by only three-and-a-half Fahrenheit degrees, which is less than 1 percent of the atmosphere's total number of degrees Fahrenheit above absolute zero.

Earth back into space without being absorbed at the surface or in the atmosphere is not subject to GHG mechanisms, so an increase in albedo generally represents a negative value for RF. For example, snow and ice on the earth's surface increase albedo and act to reduce RF. However, while clouds increase albedo and reflect inbound radiation, they also act as a blanket to retain any *outgoing* radiation. Various kinds of particles, when suspended in the atmosphere, can either increase or decrease albedo, depending on their size and color, and can also act as nuclei for condensing water vapor into clouds and thereby alter the net result of their effect.

All of these RF factors and their interactions have been studied in great depth. A useful update of their values was included in an article by Lee et al. (2009), which provided much of the data for the paragraphs that follow. In this discussion we are concerned with two things: estimates of the net effect of each of the various substances and—just as important—the reliability of those estimates.

Greenhouse gases first: the important ones for our purposes here are CO_2, ozone (O_3), and methane (CH_4). But initially, let us talk about water (H_2O). Water vapor is a hugely important greenhouse gas but its effect is limited by the low atmospheric residency time of any excess: for any given global average temperature, a certain amount of water vapor will remain in the atmosphere, but any amount above that will condense out as precipitation. Since anything that does not change the current atmospheric balance will not change the current temperature, water vapor's tendency to manage itself at a fairly constant level will normally discount its importance as a GHG RF factor.

However, water vapor has other more specifically aviation-related effects. A by-product of fossil fuel combustion, water vapor inserted directly into the atmosphere at higher altitudes may not alter its total amount in the atmosphere (thanks to the stabilizing effect of precipitation). On the other hand, it may change the amount that normally occurs in the form of high-altitude cloud. As just mentioned, this turns water into a substance that increases albedo, producing a negative RF. However, if that cloud traps more radiation (which tends to happen at night) it *increases* RF. The sum and balance of these effects occur in a way that is not well understood. The mechanisms can actually be observed: When the conditions are appropriate, the water vapor in jet efflux condenses to form the familiar white lines that sometimes mark the sky as an aircraft passes. These condensation trails (contrails) are estimated—again with error bars—to have a substantial warming effect on the atmosphere's heat balance. But the deposition of water vapor also leads to subtle types of cloud formation, referred to as aviation-induced cirrus (AIC). AIC can result when contrails fail to dissipate quickly and instead grow, but can also arise spontaneously from water vapor that had been deposited at altitude. Radiative forcing is measured in milliwatts per square meter ($mW\ m^{-2}$). The net of all RF effects of high altitude deposition of water vapor from aircraft is currently estimated to be a positive (harmful) value of about 0.9 $mW\ m^{-2}$ (ranging from about 0.3 $mW\ m^{-2}$ to about 1.4 $mW\ m^{-2}$ (Wilcox, Shine and Hoskins 2012). These values are quite different from estimates made in earlier years, illustrating that scientists are still learning about the effects of specific substances.

Now let us turn to ozone and methane. These gases are not actually emitted by aircraft, but oxides of nitrogen are. The mono-nitrogen oxides—nitric oxide (NO) and nitrogen dioxide (NO_2), collectively referred to as NO_x or NOX—interact with atmospheric oxygen to make ozone, which is a potent GHG. Methane, always found in the atmosphere (due to agricultural activity, for example), is an even more potent GHG. However, ozone breaks down the more powerful methane, in an immensely complex dynamic that is not well understood at all. Netting out the estimated positive and negative RF effects of this chemical chain of events suggests a significant increase in the aviation emissions total RF but there is room for some doubt.

The situation becomes yet more difficult when we try to assess the net effects of aerosols and cloud formation. Aircraft emit carbon black (soot) together with sulfate aerosol. These materials affect atmospheric albedo directly and also serve as condensation nuclei in the process of cloud formation. Their effects are thought to be small but the error bars are substantial.

How is Aviation's Effect Reported?

Radiative forcing is generally categorized in terms of carbon and non-carbon effects, which is really understood to mean CO_2 and non-CO_2 effects. Understanding the way in which some of this information is reported in the literature is hardly less confusing than the climate mechanisms themselves. As mentioned, non-carbon effects and those from methane and ozone are difficult to assess, but the other non-carbon effect that stems from the concept of AIC seems to be the most problematic: the reporting criterion tends to quantify the RF of all aircraft-caused cloudiness and therefore subsumes contrail formation. This, in itself, is a complex assumption.

The foregoing descriptions are necessarily very general and do not attempt to explain the reasons for any uncertainty. They are presented only to illustrate how difficult it is to be precise and to acknowledge that current estimates may understate the problem. Of course the estimates may also overstate the problem, but that seems less likely and would probably not be by nearly as much.

In light of all of this and reflecting different levels of confidence in the assessment of mechanisms, aviation RF is often presented in terms of

- CO_2 ('carbon') effects;
- all effects excluding AIC; and
- all effects including AIC.

The numbers presented by Lee et al. (2009) are unsettling: in 2005, the aviation CO_2-only mean RF estimate worked out to be about 1.6 percent of total anthropogenic RF. But adding the other effects mentioned (except AIC) more than doubles aviation's share, to 3.5 percent; including AIC yields an estimate of more than triple the CO_2 figure: about 5 percent. This is already worrying, but working within the error bars, appropriately summed according to what can and cannot be netted, an even higher number could result.[4]

It is extremely important to bear this combination of numbers in mind when looking at things like emissions reduction goals for the aviation sector; if science confirms the mean values of all effects, aviation's contribution to global warming becomes more significant than currently estimated. Even rendering aviation completely carbon-free would only reduce its RF profile by less than one third when we account for the remaining high figures for non-CO_2 effects and aviation-induced cirrus. If aviation produces these sorts of numbers in calculations of its current share of the problem, what can be anticipated in the longer term?

Before we move on to the future, and given what we have just said about the difficulty of establishing precise assessments, we should not entertain any notion that the scientists have no idea what is going on. For instance, it is worth a few more words on the reliability of the more fundamental CO_2 information. We have to accept that there can be errors, but we also have to know that on some points, there is a fair degree of consensus. There is certainly disagreement about the possible range for the CO_2 data, but the disagreement is in the bounds of small numbers. The United States Government Accountability Office (GAO) prepares reports that review and summarize data from a variety of sources. For example, *Aviation and Climate Change* (Government Accountability Office (US) 2009) cites an analysis prepared by the authoritative United Nations Intergovernmental Panel on Climate Change (IPCC)[5] indicating that aviation's current share of CO_2 emissions constitutes 2 percent of the world's total. We have encountered figures of 1.6 percent or 2–3 percent already, and although these numbers vary quite a bit as a proportion of each other, their values, in absolute terms, are proximate and small. The same 2009 GAO report cites US Environmental Protection Agency (EPA) data[6] as claiming that domestic aviation emissions account for about 3 percent of the United States' total. The IPCC and EPA estimates are thus fairly consistent—aviation's share of CO_2 production is 2 percent of the global total, and is closer to 3 percent of the national total. That seems about right in a developed country where per capita air travel can be presumed to be high. We are concerned with global totals, so I will proceed on the basis that at least this assessment is sufficiently close to the actual: about 2 percent of current global anthropogenic CO_2 is from aircraft.

In summary, we have relative agreement on the CO_2 figures and an open question on exactly how much the other effects would raise aviation's contribution

---

4   Total aviation RF or particular RF mechanisms are sometimes described as multiplying the raw carbon number. Atmospheric CO_2 and the RF that it produces act as a reference value, and other effects are expressed as CO_2 equivalent (CO_2e).

5   For the full IPCC source documents, see Intergovernmental Panel on Climate Change (2007b).

6   For the EPA source document, see United States Environmental Protection Agency (2009).

to global warming. We next take these uncertain totals and think about their coming evolution.

The Size of the Sector: The Raw Numbers Set to Grow

Since the air travel business provides an important infrastructural element in the world's economy, aviation is anticipated to grow as the economy grows. But the global economy does not merely grow in size without also growing in complexity and in the level of interaction between regions and nations. The fact that air travel takes place to support that degree of integration implies that aviation must grow a bit faster than the larger economy. Leisure air travel, already an enormous part of the business, is accelerated by progress in economic development (along with increases in efficiency and reductions in the cost of flying). Consequently, as the economy grows and a country's gross domestic product (GDP) rises, more and more people will want to travel for pleasure, leading to further expansion of air travel that will again tend to exceed general economic growth.

This notion is supported in industry history and forecasts. The Commercial Airplane division of the Boeing Company predicts that in the 2012–2032 time frame, growth in revenue passenger-kilometers (RPK) (estimated at 5 percent) will outpace growth in GDP (estimated at 3.2 percent), a ratio of 1.6 to 1 (Boeing Company 2013). In the shorter term to 2014–2015, a July 2013 press release from the International Civil Aviation Organization (ICAO) foresees a 5.9–6.3 percent increase in world air traffic, with a GDP increase of 4.0–4.5 percent per annum, a ratio of about 1.4 to 1 (International Civil Aviation Organization 2013c). Looking slightly further ahead to 2017, the International Air Transport Association (IATA) forecasts RPK growth rates of 6.4 percent (system-wide) and 7.2 percent (international alone) (International Air Transport Association 2014). This RPK–GDP relationship is not new: Airbus SAS, in its 2012 *Global Market Forecast*, notes that over the past 40 years, when GDP increases by 1 percent, air traffic can be expected to increase by 1.3 percent (Airbus SAS 2012). All of these figures support the notion that air transport is not merely a derived demand; aviation seems to be participating in the development of growth rather than simply responding to it.

A conservative industry growth rate of 5 percent per annum yields a doubling every 15 years, and would therefore compound into an approximate sixfold increase by 2050. Barring actions to reduce either the energy intensity of flying (for example, by consuming less fuel) or the emissions intensity of flight energy (for example, by consuming fuel that embodies less carbon), and ignoring the possibility that airspace or facility constraints would moderate growth at some point, the global warming impact of aircraft will track right along with that high rate.

What is Aviation's Future Share?

Studying forecast increases in aviation's absolute level of emissions tells us how much bigger aviation's RF numbers could be, but let us understand that policies, the industry's response to them, and support for them will be strongly affected by the 'aviation versus overall GDP' discussion. So we must go back to percentages. We have seen that aviation is growing faster than the overall economy. It would be useful to know its future share of the total effect. The share of the global warming account attributable to any particular activity or state in the future is not simply dependent upon how successful it is in reducing emissions, but also how successful everyone else is. I am going to make brief reference to a study that was designed to posit different possible paths of world economic development under various assumptions regarding economic and commercial policy and emissions reduction policy, so that the shape of growth in total future emissions could be envisioned.

Special Report on Emissions Scenarios (SRES)

The Intergovernmental Panel on Climate Change (IPCC) is a creature of two United Nations organizations, the United Nations Environmental Programme (UNEP) and the World Meteorological Organization (WMO). Following the Rio Earth Summit of 1992[7] and the global commitment to the United Nations Framework Convention on Climate Change (UNFCCC), the IPCC was established as a sort of operational unit for the study of global warming and its effects; it has the key role in managing the world's effort to collate and communicate the nature, scale, and evolution of the climate crisis, and ways to address it. As part of its mandate, the IPCC produces *Assessment Reports* every few years and occasionally *Special Reports* on particular topics, which are used as reference materials by many experts. Early on, it became apparent that in order for climate scientists to be able to create relevant computer models that could predict the general behavior of Earth's changing climate, there had to be some formal, documented assumptions about how emissions activities might proceed in the future. Based on data collected for the Rio Summit, in 1992 IPCC published a series of emissions scenarios, often referred to as 'IS92' (Leggett, Pepper and Swart 1992), which were then re-examined and expanded in 1996[8] to result in the *Special Report on Emissions Scenarios (SRES)* (Nakicenovic and Swart 2000).

---

7 See Heinrich Böll Foundation (2003) for a brief report on the 1992 Rio Earth Summit.

8 A major 2007 IPCC report concludes that the 1996 scenarios projections are still valid: 'There is high agreement and much evidence that with current climate change mitigation policies and related sustainable development practices, global GHG emissions will continue to grow over the next few decades. Baseline emissions scenarios published

The production of these scenarios certainly represents an impressive application of complex simulations of economics principles joined with detailed analyses of the implications of different sorts of policies and social and political developments. But they are, by their nature, tentative. Their value here is not in the specific pictures that they describe but rather that they produce a credible range that allows us to see how aviation might fit in. Presented in the following paragraphs is a discussion of a little of what the *SRES* says about how global amounts of emissions might evolve.

Four scenario family 'storylines' are developed, each with sub-scenarios examining the effects of variations in any of the scenario-driving forces—population, economic and social development, energy and technology, agricultural and other emissions, and policy. In general, what the SRES refers to as A1 and A2 scenarios reflect a stronger focus on economic development, while B1 and B2 scenarios place more emphasis on environmental and social concerns. (A1 might be called a moderate worst-case scenario, and is the type represented in much of the current literature.)

The complexity of the computations and the detailed analysis of the results is beyond the scope of this chapter, but the *SRES* summaries, graphs, and tables provide a telling outline of the range of possible family-based scenario outcomes: global energy-use CO_2 emissions to 2050 nearly triple with the A scenarios, and nearly double with the Bs (Nakicenovic and Swart 2000).

In predicting growth to 2050, however, *any* possible global emissions scenario rates need to be set against the previously predicted *sixfold* growth in the aviation industry and its associated emissions. And even in a moderate worst-case scenario, whatever the rest of the world may do to reduce its emissions rate, aviation's still outstrips it *if the aviation industry does not take action*.[9] The last is an important qualification, because this book is all about doing something.

These scenarios are designed to give general shape to our understanding of emissions growth, though they are hypothetical and cannot guarantee accurate forecasts of what will actually happen. Nonetheless, the problem is clear: in the foreseeable future, even if little is done to reduce emissions in other kinds of economic activity in the world, aviation's share of contribution rises. If the rest of the world economy works really hard at reducing emissions, aviation's relative share rises even higher. Hardly surprising: we could tell from the figures mentioned earlier that this was the way things were evolving. But this is all extremely important from the point of view of managing the industry's fortunes, and indeed its capacity to serve its infrastructural development role.

---

since the IPCC Special Report on Emissions Scenarios (SRES 2000) are comparable in range to those presented in SRES.' (Intergovernmental Panel on Climate Change 2007a).

9   Furthermore, bear in mind that we are discussing *rates*: an emissions *growth* rate of zero means that we would still be pumping the same amount of GHG into the air in 2050 as we are today. And a *current* emissions rate of zero means that we are left with the accumulated quantity extant.

If we now understand both a little about the particularly complex ways in which aviation meddles with the atmosphere, and if we then also understand how the effects will grow in absolute and relative terms, we can start to form ideas about how and how much aviation *should* reduce its impact, and shift our attention onto the substance of what can be done. Looking ahead, we find that there is a lot to be done; subsequent chapters will each deal with different aspects.

In examining the emissions question for any particular activity, three considerations drive the total: the extent of the activity, the energy intensity of the activity, and the emissions intensity of the energy employed in the activity. This is a formulation to which we will return again and again in many different contexts. The extent and growth of aviation activity has already been mentioned. Using fuel more efficiently would reduce the energy intensity of aviation activity, so what can be done in this sphere? We may not want to curtail flying at this point, but we do not yet have much low-emissions fuel, so the next chapter will look at the very first thing that needs to be considered: possibilities for making better use of the energy. Discussions of emissions intensity of energy will follow later.

Chapter 2
Aviation's Energy Predicament: Reducing the Wingprint

Fuel Efficiency

This is the first part of the discussion about steps to reduce aviation's global warming effect: improving the way that we use the fuel that we burn. When we talk about reducing the fuel consumption of air travel, we have a ready set of factors that we imagine will do the job. Engineers know them: maximizing airframe aerodynamic performance, reducing aircraft weight, increasing propulsion efficiency, streamlining aircraft operating procedures, improving air traffic management (ATM) systems and procedures, and optimizing load factors. In fact, these have always been goals for commercial aviation. There is evidence that aviation has improved in every one of these areas and has achieved admirable results in reducing energy intensity of air operations during the years that mass air travel has been available. It is not just concern about the environment; these are commercially important matters—fuel costs money. And even if fuel were free, it still takes up volume and weight capacities that could be allocated to payload. For much longer than climate change has been a public concern, the cost of carrying and burning fuel has encouraged airlines and manufacturers of aircraft and engines to whittle constantly at the amount of fuel it takes to move one passenger one kilometer toward destination, and great strides have been made. Revenue-specific fuel consumption is about one fourth of what it used to be: at an International Civil Aviation Organization (ICAO) Workshop held in February 2009, a Vice President and the Chief Economist at Airlines For America (A4A), John P. Heimlich, presented figures for the United States (US) illustrating a 27 percent gain in revenue ton-mile per gallon between the years 2001 and 2007 (Heimlich 2009). A few months later, at a November 2009 ICAO Conference, an International Coordinating Council of Aerospace Industries Associations (ICCAIA) overview indicated that overall, fuel burn had been reduced by 70 percent between the 1960s and the 1990s, and the trend had continued to a fourfold increase in commercial aviation fuel efficiency between 1960 and 2004 (International Coordinating Council of Aerospace Industries Associations 2009). This is an average compounding improvement rate of just over 4 per cent. Building on this progress, at the 37th Session of the ICAO Assembly in 2010, members resolved to achieve an annual 2 percent improvement in fuel efficiency to at least 2020 and perhaps even to 2050 (International Civil Aviation Organization 2010).

Airspace

However, the very next slide in Heimlich's 2009 presentation was less encouraging: in 1968, the block time (total flight time gate to gate) between New York and Washington was 60 minutes, but in 2008 it was nearer 75 minutes. These data were perhaps offered in a context different from the point that I wish to make here, but they give rise to an interesting question: When we look at the assortment of factors that result in greater efficiency, do they all necessarily lead to reductions in emissions? Great investments were made in ATM improvements in the US in those intervening 40 years, but despite all of them, aircraft of a similar speed took 25 percent longer to cover the same distance. The efficiency gains were not trivial, nor were the initiatives undertaken to achieve them: the introduction or evolution of new communications tools, radar, computers, transponders, display technology, task automation, and airspace structure. Still, in this traffic-dense northeastern airspace, congestion slowed airplanes down. What was achieved? The numbers of flights grew enormously, and if improvements had not been made in other areas, emissions on a unit capacity basis would have gone up. But we need them to go down.

There are good reasons for increasing the capacity of the airspace system but reducing emissions is not going to be one of them, unless some stabilization of levels of traffic volume can be attained or the efficiency with which we use that capacity soars. In some parts of the world, very large increases in traffic can be anticipated. In mature market regions, progress in new airspace management initiatives such as Single European Skies ATM Research (SESAR) in Europe or NextGen in the US may outpace traffic growth enough that we will see per-flight reductions in emissions. For example, in the initial drafts of its Master Plan, the SESAR initiative envisaged a 73 percent increase in airspace and airport capacity (over 2004 levels) by 2020, an eventual threefold increase in European ATM capacity, and a 10 percent reduction per flight in environmental impact (Single European Sky ATM Research Consortium 2009). But even though European air traffic has not grown quite as strongly as predicted since the introduction of this plan, airspace performance targets have been substantially lowered (Single European Sky ATM Research Consortium 2012), and pressure to deliver results is increasing (European Union 2012). Nonetheless, if indeed capacity and efficiency grow more quickly than traffic, there is cause for hope.

It may be difficult to know exactly how airspace efficiency gains will play out in the European Union (EU) and in North America, whose economies are relatively mature in terms of level of development and access to travel capacity. But much of the immense amount of foreseen growth in the air industry will occur in regions of the world that are not so developed, and part of the answer to the larger question about the effect of ATM modernization lies in this fact. Perhaps such regions will not be capacity-constrained if they have access to modern ATM technologies—but the picture is worrying. Much of the increase in domestic and regional air travel in Asia will be in airspace between centers that have relatively little traffic now; in an article on the introduction of the low-cost carrier model in China, Liang and James (2009) noted that while the average US resident makes over two air trips in a year, only one person in 13 travels by air in China. On the other hand, the article included statistics from the Civil Aviation Authority of China indicating that growth in air travel was predicted to run at an astonishing 17.5 percent, which is a doubling every five years or so. More recent reports now foresee a more moderate rate of growth (CAPA–Centre for Aviation 2012), but one still wonders if Asia can grow without straining ATM capacity and incurring congestion-related fuel burn and emissions growth even on a capacity-specific or revenue passenger-kilometer (RPK) basis. When we think about already more congested regions like India, the picture becomes even less rosy. Some will hope that these possible saturation issues will serve to moderate traffic growth, but so far they have not.

I certainly do *not* take the position that we should not support improvement of our ATM capability; I only say that it is difficult to quantify the effect on emissions. Undertaking an analysis is beyond the scope of this book but I raise the questions here as a caution that we should be prudent when estimating the role that this modernization will play in reducing emissions.

Operations

One of the options mentioned as offering opportunity to reduce emissions is improvement in aircraft operating procedures: flights can be planned more optimally, and pilots can handle the aircraft and management of the flight in a way that reduces fuel burn. While true, this is also somewhat simplistic. Yes, it is always possible to find a slightly better solution to a strategic or tactical flight problem, but in this regard airlines and their pilots are now gnawing on some pretty well-chewed bones. I can say from personal experience that reductions in fuel burn due to improved operational procedures, having been pursued assiduously for decades, are getting progressively harder to achieve. Gains now are quite limited. Additionally, many aspects of efficient flight operations are constrained by airspace management systems about which we have just spoken, so the answers are not very clear or encouraging here either. Still, they should be pursued and will yield something. Certainly, the necessary improvements in ATM combined with the better aircraft that we will discuss in the next section, can only be capitalized upon if airlines and their pilots exploit them properly.

Aircraft and Engine Improvements

We can confidently state that increasing the efficiency of aircraft and their engines does reduce fuel consumption. Aircraft manufacturers have always had this fact fixed firmly in view. A few retrofits can enable existing aircraft to perform better; for example, aerodynamic devices such as 'winglets' or 'sharklets'—the drag-reducing fins mounted at wing tips—can be installed or updated. But these retrofits, while extremely worthwhile, only have a limited effect in reducing fuel burn; the greater need is to develop new ways of making aircraft so that substantial improvements can be realized as fleets are renewed. And one of the simplest ways, though certainly not the easiest, is to reduce weight, especially for long-distance missions. Large long-haul airplanes weigh hundreds of tonnes. But if the weight of such an aircraft is reduced by just one tonne, the fuel that it burns between each takeoff and landing will be reduced by hundreds of kilograms. By one calculation, a 1 percent reduction in landing weight could result in fuel savings of 0.75 to 1 percent, depending on the type of engine (Boeing Company 2004). I know through personal observation that an airliner deployed on a regular, very long-haul run— between Toronto and Hong Kong, for example—will reduce its fuel burn by more than one half kilogram for every kilogram reduction in landing weight of the aircraft each and every time it flies that route.

New materials are enabling aircraft manufacturers to accomplish significant weight reductions; empty weights of aircraft have been coming down steadily as new designs incorporating new substances are introduced into the market. Though much-heralded carbon fiber composites cannot be called new, they are now really gaining ground. Caution and stringency in the evolution of certification criteria have meant that their incorporation into airframe designs has been slow and careful—but steady.

Let us look at some other airframe and propulsion improvements that might appear in the coming years. What technologies can we look for, and what results can we expect them to produce? Green (2009) equips us with a very useful summary of some of the prospects for reducing fuel burn and emissions and I make reference to those ideas in the paragraphs that follow.

Weight is mentioned repeatedly of course, but the first point specifically detailed in this regard is that designing aircraft to be large and strong enough to carry the fuel required to make long flights and then loading that amount of fuel is far less fuel-efficient than simply using slightly smaller aircraft to carry the same payload over shorter stage lengths, and refueling. As a corollary, due to requirements of mission flexibility, long-range aircraft are often used on shorter flights, which is even less fuel-efficient. While the airlines need to be allowed to design their own product, it is useful for everyone to consider the costs associated with particular types of flying.

Moving beyond weight, propulsion technology is seen as another area that could yield results. The total efficiency of a turbo-fan engine's operation is made up of a number of fairly discrete elements. These include thermal efficiency (extracting heat energy from the fuel), transfer efficiency (turning heat energy into mechanical energy), and propulsive efficiency (turning the mechanical energy into thrust). The first two now approach their theoretical limits, but design concepts that might reduce an engine's pressure ratio and increase core temperature may provide for more effective use of fuel.[1] Furthermore, the efficiency of the fan itself seems to be an object of encouraging work. The large front-mounted impeller (fan) on a current engine can be altered and moved to a location outside the engine housing (nacelle) where it can spin un-shrouded, like an old-fashioned propeller—actually two old-fashioned propellers, one directly behind the other, rotating in opposite directions. Each propeller could have several blades instead of the familiar two or three or four. This open rotor design is more complex than the current one; it would require a gearbox to make the rotors spin at appropriate speeds, but it may well offer better performance. One challenge will be to ensure that the exposed fan does not make too much noise.

Examining the aircraft itself for its potential gains in aerodynamic efficiencies is basically an exercise in improving the ratio of lift to drag (L/D). Drag is generally made up of two types of resistance: the kind that is generated by simply trying to push the airplane forward (you experience the same thing on your bicycle) and the resistance that results as a byproduct of generating the lift that holds the airplane up. The wing is the part of the airplane that generates lift; the entire aircraft generates drag. So one way to think of an airplane is simply as a wing (which is the useful part) burdened with other necessary but aerodynamically inconvenient appendages. Wings are now so cleverly designed that the drag that they induce in holding up the aircraft's weight does not provide very much potential for reduction, but there is some. There is also room for improvement in smoothing the flow of air over the wing surface. Technologies to increase laminar flow can accomplish this; they have been demonstrated and do offer promise. Passive improvement of laminar flow can be achieved by tailoring wing shape and sleekness; active systems use a porous surface and suction to draw non-laminar, turbulent flow down into the wing.

Increasing the wingspan also improves the L/D ratio, but this results in a slower airplane. The classic aft 'sweep' of modern jet aircraft wings has an aerodynamic purpose in allowing more of the wing to remain efficient at higher speeds, but such wings are difficult to lengthen without incurring penalties in weight and thickness. However, even if reducing sweep and lengthening the wing restricts aircraft speed, this still might be worth considering, especially for aircraft that are intended for shorter missions.

Now we come to the matter of appendages. If the payload-carrying part of the aircraft—the fuselage, or body—can be eliminated and this space incorporated into the wing itself, then the entire aircraft becomes part of the wing and there is no non-lift-generating remainder to create mere drag. In the extreme expression of this idea, the aircraft is simply a wing. A compromise is a blended wing/body design.

Taken together, some of these ideas offer a substantial saving; for example, it is estimated that a full laminar flow 'flying wing' could burn less than half the fuel

---

1　Engine designs are addressed in more detail by Kyprianidis et al. (2011).

that is consumed by aircraft using current configuration and technology (Green 2009). That is truly impressive.

Mission

Another point made in this research (a point that relates again to mission and has long been known) is that a disproportionate amount of fuel is burned on short- to medium-range flights. Just as it costs fuel to design and operate aircraft for extremely long missions, the same is true for airplanes that do short hops. There might be ways to reduce the amount of this kind of flying, but perhaps aircraft typically undertaking short flights should be prioritized for incorporation of new design features. It may well be that some of the more dramatic design changes, such as the flying wing, will not lend themselves to replacements for smaller aircraft like regional jets from Bombardier and Embraer, or the Airbus 320 and Boeing 737 families of airplanes that populate short- and medium-length routes, but the need for improvement at this mission scale is most important.

In examining the sorts of performance improvements that research is targeting or envisioning, a global, fleet-wide, capacity-specific fuel consumption and emissions improvement of approximately 65 to 70 percent seems feasible (Green 2009). When we consider cycle times in fleet renewal, this is not out of line with other estimates. Research at the National Aeronautics and Space Administration (NASA) explores some of the same ideas and a few more in its Subsonic Fixed Wing Project (a collaboration with industry and academia), and is working toward an improvement of more than 70 percent over 2008 levels of fuel consumption for aircraft entering service in 2030 (National Aeronautics and Space Administration 2010).[2] The *Vision 2020* undertaking of the Advisory Council for Aeronautics Research in Europe (ACARE) imagined improvements on the same order, with a 50 percent reduction in CO_2 emissions between baseline 2000 and service entry of 2020 (Advisory Council for Aeronautics Research in Europe 2010). These studies and initiatives also point to similar orders of decrease in mono-nitrogen oxide (NO_x) emissions. *FlightPath 2050*, ACARE's current initiative, incorporates recent developments in new technologies and alternative fuels, and aims for even greater emissions reductions than the now-superseded *Vision 2020* (Advisory Council for Aviation Research and Innovation in Europe 2012).

---

2   A 2009 analysis of trends since 1960 indicated that improvements in efficiency had declined slightly in recent years due to the relatively low price of jet fuel and the persistence of older aircraft in service (Rutherford and Zeinali 2009).

Is It Enough?

So all of these projects foresee really great gains. But we have to determine how much these rather impressive anticipated improvements in efficiency might help us in reducing emissions. Let us see where the numbers take us over time. If we were to assume that aircraft entering service in 20 years or so will have achieved a 70 percent reduction in fuel consumption on a revenue load-specific basis, the entire fleet would probably reflect that kind of progress by 2050. But recall that the aviation industry is expected to grow by about 5 percent per year during the same time period, yielding a sixfold increase. Fuel will be consumed by an ever-increasing number of aircraft and flights. Accepting that GHG emissions are directly proportional to fuel consumption, aviation's current contribution to total atmospheric GHG would still nearly double by 2050, even with the aforementioned impressive 70 percent gain in fuel efficiency. All of this effort and yet we would end up putting more greenhouse gas into the atmosphere every future year than we do *this* year.

We must also remember that an immense unknown in the discussion of global warming was the extent to which aircraft contrails and aviation-induced cirrus (AIC) might increase the climate change effect. Earlier we noted that research has assessed that contrails and AIC have significant effect, and more precise data are now being recorded. Recent studies also suggest that while contrail/AIC effects are extremely strong, they can still be addressed. Investigation conducted by Ulrich Schumann at the Deutsches Zentrum für Luft- und Raumfahrt (DLR)—the German Aerospace Center—focuses on these effects. In a presentation to the ICAO Colloquium on Aviation and Climate Change in May 2010 (Schumann 2010), and in a more detailed paper presented at a conference sponsored by the American Institute of Aeronautics and Astronautics (Schumann, Graf and Mannstein 2011), Schumann described the tools being used to study these climate mechanisms, and also showed that flight planning to minimize radiative forcing (RF) from AIC could produce dramatic reductions. It may be true that planning off-optimum altitudes and routes could result in some incremental increase in other emissions, but the net result would seem to be worthwhile. Let us hope this turns out to be correct and that something feasible is produced in that regard. Because if we think of the CO_2 and other RF effects all taken together, they would have to be cut in half just to keep us at the same level of effect as the industry produces right now.

But, thinking about just the CO_2 for a moment: an increase in the absolute amount of emissions no matter what we do? Not acceptable; we need emissions to fall, not rise. In fact, let's recall the previous chapter's discussion of the *Special Report on Emissions Scenarios*, which predicted a possible doubling or even tripling of global emissions by 2050 in a moderate worst case. Over that time frame, the supposed best possible effort of aviation alone, resulting in industry emissions growth to nearly 200 percent or more of current, leaves it looking relatively better only if the rest of the world fails miserably.

There are lots of ways of presenting and hypothesizing about the data that we have gathered, and our assumptions about improvements in efficiency can take other elements into account as well—for instance, I have not talked about room for improving load factors. But the important point is that no one—even those supporting the air industries—contends that efficiency alone is going to be an answer to the challenge.

A *New* New Aircraft Design

Before we close the chapter we must answer the question that unavoidably arises: is there a different, more dramatic technology or suite of technologies that completely departs from the way in which we power aircraft now? Current aircraft burn a conventional type of liquid fuel. Can we create a new fleet that uses renewable liquid hydrogen (LH_2) or perhaps even a practical and safe form of nuclear (!) energy? The short answer is yes, we should and we will change the global fleet in some fundamental way, eventually. The scale of effort that is required to address the open environmental questions here is unprecedented, and calls for an examination of even the most extreme and extraordinary possibilities. It seems inevitable that exotic technologies will have a role to play in air travel's future, but let me make this point early on: Anything that we discuss as a replacement for current technologies will have a long gestation. Aircraft based on completely new technologies must first be imagined, the ideas researched and developed, specific designs created and compared, financial commitment to option(s) obtained, construction of working examples completed, and the initial line versions of the design manufactured. Each step in that evolution can take many years in an industry that must meet standards of high performance, efficiency, and safety that would beggar the imagination of anyone not familiar with the way that aviation and its regulation function. Additionally, once we have new aircraft entering the fleet, it will be decades before the 'old' technology aircraft are amortized and the exotic newcomers become the majority.

As an example, look at the recent experience of bringing two aircraft to market; both are new models, though based on comparatively traditional technologies. While the Boeing 787 (B-787) and the Airbus 380 (A380) are certainly wonderful expressions of progress in aeronautical design science, they are not examples of the kind of wholesale technology change that we are considering here. Still, it is worth noting just how long even such relatively conventional projects can take.

The B-787 (called the 7E7 in early stages of the project) was an idea that gained corporate project commitment in late 2002 upon cancellation of the proposed Boeing Sonic Cruiser. The options being considered at that date (*including* the B-787 and Sonic Cruiser) were a product of research streams that began to take form in the late 1990s.[3] The B-787 entered service in late 2011 after several delays. Similarly, development plans for Airbus's double-deck A380 (3XX at the concept stage) had been made and described in the media in the mid-1990s (Sweetman 1994), with a project launch slated for 1997 or 1998 and delivery anticipated in 2003 or 2004. It too was late, entering service in 2008.

Even conventional aircraft have taken ten years and more to get from a firm concept to line operations; an altogether new sort of aircraft would face even more daunting development times. It is reasonable to expect that the more fundamental preliminary research into new and daring technologies might be two or three times what is described above, possibly 20 years of concept research and development before the manufacturers could even begin the ten or more years of work required to bring those new-technology aircraft to the flying world.

So the question becomes not whether we can do something entirely different, but rather how we are to get to that point in a sustainable manner *even if* we are certain that we will ultimately develop completely new ways of powering aircraft.

I do not mean to downplay the air industry's capabilities. In fact, I think that its track record of gains in safety and efficiency is an example to the whole world of just how well it can perform when it dedicates itself to improvement. My point is that while it must try in every way to get better in the both the short and the long term, certain strategies are going to have limited or distant results. Critics may claim that aircraft manufacturers are not trying hard enough, or that airline customers are not demanding enough, but my feeling is that accelerating the pace of development of new technology would require the kind of whole-society support that we usually only see in time of war, for example. An entirely new kind of aircraft could be introduced *a little* more quickly, but that would be a political decision within each of the few individual nations where aircraft manufacture resides.

If this is the case, we must accept that we have to solve the carbon emissions problem in terms that accommodate the use of an expanding fleet that is in key respects similar to the airplanes that we operate now. We will make the new fleet efficient, and we will use it efficiently as well. But we must start thinking not merely about the energy intensity of flight but rather shift the discussion toward the emissions intensity of that energy. We must start looking at the fuel itself.

If we are stuck with familiar sorts of aircraft for the time being, we must first understand a little bit more about the fuel that they burn. Then, in later chapters, we can start examining the project of reducing the fuel's carbon content and the things that will affect that prospect.

---

3   A *Flight International* article (republished online) describes the transition from the Sonic Cruiser to the 7E7 and how the concepts were being plotted in parallel in that time frame (Norris 2003).

Chapter 3
Aviation's Energy Predicament:
No Fuel Like an Old Fuel

Let's Use a Different, 'Renewable' Fuel

When one starts to think about ways of making ordinary, current-technology aviation more sustainable while some complete alteration of the underlying technology percolates through the research and development (R&D) establishment, there is value in asking the obvious question about our fleet's ability to simply burn something else. 'I understand that aircraft must burn liquid fuels. I understand that they are going to be doing that for a long time. But I do not understand why it has to be kerosene. Why not just load bio-ethanol or some other kind of renewable fuel into the beasts and head off into the wild blue? Maybe there is a kind of liquid that is renewable and sustainable and that grows on trees—there for the picking. The trees would gather carbon dioxide (CO_2) out of the air in order to thrive and produce the oil-bearing fruit. Then, when we burned the oil in airplanes, the CO_2 released by combustion would be gathered in again by the trees to create more oil. Wonderful. What's the problem?' The answer to these questions is fairly simple in the minds of some in the engine and airframe contingent, but it is worth taking the time to work it through in sketch form because it is at the core of strategy development as regards emissions. It turns out that a lot of things are put in play by such new-fuel propositions. (We will get to the difficulty associated with carbon life-cycle analysis and the matter of true sustainability later.) First, let us ask what is the more immediate technical problem with pursuing some different fuel.

In general, the core issue is that intensive airplane and engine development work would have to be undertaken just so that aircraft could use some (any) new sort of fuel material, which would mean facing many of the same aircraft technology challenges that were described in the previous chapter. In essence, only an airplane with a new power plant and fuel system design could burn a wholly new fuel. In other transportation settings, this is not the case; cars and buses can be designed to burn a few different substances, and it has to be admitted that airplanes have also sometimes been built with a certain fuel flexibility in mind, as we will discuss. But for now we are basically stuck—why?

Aviation technology evolved over a long time, and in some cases various conclusions and strategic decisions (if they can even be called 'decisions') occurred at different stages for reasons that do not now pertain. This wandering down a particular path in ignorance of considerations that would come to light in the future now results in a few extremely difficult energy-source constraints for air travel. Nothing about this is simple. But as frustrating and exhausting to deal with as it appears, breaking down the puzzle piece by piece reveals it to be less intractable than one would imagine. And *most* of the reason that jet aircraft burn kerosene has had a very logical path.

Kerosene: the Right Fuel at the Right Time

It is useful to start with a very brief account of how we wound up with kerosene in the first place. A fascinating little piece of aviation's archive: The first jet-powered flight in the United Kingdom (UK) took place on May 15, 1941—the Gloster E 28/39, which had been built to use and test the Whittle W1 engine.[1] Afterward, the pilot, Flight Lieutenant P. E. G. Sayer, filed a standard test flight report. Under 'Engine', Sayer had entered 'Whittle Supercharger Type W.1'. Under 'Weights Carried', the word 'Petrol' was struck out and 'Parrafin 50 galls' [*sic*] entered (Taylor and Munson 1973). Even at that early time, kerosene ('parrafin') was the fuel in use. In Germany, turbojet-powered aircraft had been flying for some time but they were also powered by kerosene (Gunston 2006). The story of why it was so then and continues to be so now is a story of practicalities.

The gas turbine engine was not by any means a novel concept, but a workable scheme waited for Frank Whittle and then Hans-Joachim Pabst von Ohain, who worked independently and developed the engine separately (Boyne 2006). The motivation was speed and power; a conventional propeller loses efficiency at high airspeeds and a reciprocating engine is quite heavy for the amount of work it can do. A gas turbine engine was seen as the solution to many problems.

Interestingly, a jet[2] engine can be designed and set up to operate on virtually any sort of liquid fuel. One very important reason that kerosene was chosen in the early days is that gasoline was a strategically important material in the period leading up to and throughout the Second World War. In times of extremely rapid technological innovation—which that certainly was—an irony exists: innovators are sought out and encouraged, but sometimes the flood of creative effort generates new ideas at a pace that exceeds society's established capacity to assess and exploit them. Individual discoveries must then compete for attention; those that do not attract enough interest will not receive the resources they need in order to develop further. This was, to a certain extent, the case for gas turbine development in the UK, though not so much in Germany. No one particularly cared whether researchers burned up kerosene in weird experimental engine prototypes, but they could not be allowed to use precious gasoline. Even in Germany, where von Ohain enjoyed industry support from the aircraft manufacturer Ernst Henkel, the strategic importance of gasoline contributed to the selection of kerosene there as well (Gunston 2006).

---

1 For more information, see Grierson (1971).

2 All jet engines are gas turbines, but not all gas turbine engines are jets. In this historical context, the terms are used fairly interchangeably.

Eventually the jet engine was recognized as an important technology that would allow military (gasoline-worthy) aircraft to reach higher speeds and altitudes, but by then designers had settled upon kerosene and continued with it.

One might think, though, that once the immediate pressure of conflict was off, with sufficient time to reflect upon the true potential of jets and ways in which the technology could be optimized, some new designs would involve a different fuel. Surely this new, sophisticated, exotic, efficient, and promising technology should be able to dictate terms, and some special 'superfuel'—tailored specifically for the turbojet's needs—would be concocted for its use? Well, different fuels were proposed, tried, and used, but the fact is that rather pedestrian concerns continued to direct jet fuel's development, and it turned out that kerosene had been a pretty good choice. First, any fuel that is going to find its way into widespread use cannot be rare and expensive, it must be readily available and cheap. This meant, from the very beginning and through the later years of the turbine engine's commercial life, that the fuel would likely be some kind of petroleum distillate. The next factor was that life aboard an airplane is cramped; everything competes for weight allocation and space. The fuel therefore had to provide a satisfactorily large amount of heat energy relative to both its weight and the storage volume it used. While kerosene performs reasonably well in this regard, it does not contain as much energy as, say, gasoline. But when users started experimenting with blends of fuel that incorporated lighter distillates, they discovered other practical considerations. So-called 'wide-cut' fuels turned out to present other problems. Users (particularly military users) who had some experience with kerosene valued its relatively low volatility and consequent safety in storage and operation. That increased level of safety fit quite well with airlines' priorities. Wolveridge summarizes kerosene's suitability:

> It has been said of aviation turbine fuel that it is the ultimate commodity. The everyday reality of routine intercontinental air travel has come to pass in large measure through the universal availability of a fuel that meets the requirements of all users. That fuel is presently aviation kerosine (sometimes kerosene), supplied against a small number of internationally accepted performance specifications (Wolveridge 2000).

The specific liquids that we now call 'jet fuel' are blends of hydrocarbon distillates that fit the general category of 'kerosene'. The distillate molecules come in different sizes and shapes, but all consist of various arrangements of carbon atoms with hydrogen atoms attached. These mixtures meet more and more exacting standards of energy per unit weight, energy per unit volume, high thermal stability, low 'freeze' point, high flash point, low vapor pressure, acceptable levels of electrical conductivity and capacitance, adequate lubricity against wear and tear in pumps and valves, and some ability to maintain suppleness in the flexible fluid seals that have been designed into aircraft and engine fuel paths.[3]

While kerosene happened to satisfy certain very basic requirements during the initial period of jet engine development, it has subsequently been found ideal in other important respects, and so we realize that aircraft design choices and materials selections have been driven by and adapted to the fuel that has now been in use for such a long time.

As the user community (airlines and the military) settled into their dependence on just a few versions of kerosene jet fuel, the aircraft and engines became more precisely designed to use specific blends in an optimal way, and the fuel became more rigorously and narrowly formulated and prepared to always deliver what was required of it. Occasionally, some unique aircraft types come along (usually military aircraft with special missions) that require their own special fuel type, but generally any gas turbine-powered airplane that operates at less than perhaps two-and-a-half times the speed of sound will burn a pretty conventional blend, with a few small variations. For example, in commercial use it is sometimes desirable to use a particular blend that is more resistant to 'freezing': at high altitudes in particularly cold conditions, some paraffinic components of a regular blend can start to precipitate out in their cold-induced waxy form. In those circumstances, using a slightly different blend of fuel is preferable to planning a longer route or a lower altitude in an effort to avoid extremes of temperature. But these special blends are not very different from the more conventional type, so they are easily available and current engines are able and certified to use them.

Do We Need the Optimum?

What I have written so far about the evolution of jet fuel argues that it has come to be considered the ideal jet fuel. But not everything needs to be optimal; for sustainability's sake, perhaps we should consider burning other types of fuel even if there is a penalty. A fuel might be heavier or more volatile or non-optimal in any number of ways but maybe it is still worth paying more to be able to use it: accept a payload cost; provide heavier but more fire- or rupture-resistant fuel storage; fly at lower altitudes—whatever. At least we would be solving the carbon issue and addressing other aspects of sustainability. Alas, this is where the new-technology barriers come in. Current aircraft are not designed to be capable—and are *not* capable—of carrying or burning any sort of different fuel. No one envisioned needing a different fuel, so aircraft and engines were designed around the fuel that was being used. There is, fortunately, *some* flexibility in terms of what we can put in a current airliner's fuel system, and we will talk about that in another chapter, because it is important. But that flexibility is limited.

---

3   For details, see Aviation Fuel Quality Requirements for Jointly Operated Systems (AFQRJOS) (Joint Inspection Group 2012).

It is hard not to repeat ourselves, but different aspects of the challenge that we face show up in different contexts, and this brings us to the point of asking how difficult it would be to change the engine and fuel system design and specifications. A lot of this book will talk about novel combustibles such as different sorts of biofuels. If we assume that a specific type of liquid, perhaps a kind of vegetable oil, could be used as jet fuel (and that *is* quite possible) why do we not just build engines designed to use that liquid?

A fundamental tenet is to assure that this liquid could be produced in a way that addresses all the elements of sustainability, and perhaps this in itself might not be a problem. But we need to remember what we said about the development time for whole new airplane technologies. Development of only the engine and fuel system technologies is similarly challenging. So, the fundamental difficulty is the relative enormity of the job of revamping fuel-handling systems and engines in aircraft similar to those in current use *together with* the introduction of an additional fuel supply infrastructure. To understand this a bit better, let us expand just a bit upon the points related to the earlier discussion on development of new aircraft models.

Industry Agility

Airlines and their suppliers do not make changes capriciously. In the early 1960s the Boeing company embarked upon an effort to win a US Air Force contract to build a large transport airplane. They lost out to Lockheed, but it seemed that there might be an opportunity for some of the work that they had done in the military bid to be applied in response to pressure from the commercial side to build a new large passenger airliner. Boeing launched into this project in the mid-1960s and the B-747 'jumbo jet' entered service in 1970. As it had famously done on a few previous ventures, Boeing—one of the largest aircraft manufacturers in the world, in business since 1916—once more literally bet the company on the success of this project.

Aircraft are enormously complex and are expected to be almost perfectly safe in an operational setting that is more challenging than anything else that humanity undertakes on a routine commercial basis. The aircraft in commercial use—a collective world fleet of over 25,000 (International Civil Aviation Organization 2013a)—must daily accommodate hundreds of thousands of air travelers, lift them off the ground, accelerate them to nearly the speed of sound (say 900 kilometers per hour, maybe 1200 kilometers per hour with a tailwind), carry them to altitudes many kilometers above the Earth where there is only a wisp of air and the possibility of temperatures like −75° C, and then set them safely back on the ground hours later at perhaps 270 kilometers per hour on a runway that is sometimes slippery with ice or clogged with snow, coping with flight visibilities on departure or approach that are sometimes nearly zero. Along the way, these stalwart aircraft may also often inadvertently encounter turbulence that stresses the structure and challenges the flight control systems. They also avoid hitting obstacles such as other airplanes. Remarkably, this is all done so well that the chances of you losing your life while partaking of this kind of service are virtually nil.[4]

We know that the airplanes that are used in this activity take many years of extremely costly development and design work to create. We should think a little bit about the scale of expectations as regards return for both the manufacturers who develop airplanes and the airlines that buy them. The development investment in time and money *must* pay out over a long period. This is such a powerful reality that the manufacturing giants Airbus and Boeing now undertake new aircraft (such as the A380 and B-787 mentioned in the previous chapter) on the basis of a segmented risk-sharing development and construction process, so that it can remain economically possible to launch new projects in an atmosphere of progressively more exacting challenge brought on by increasingly stringent requirements for safety, performance, environmental benignity, and economic usefulness (Horng 2006).

As a consequence of all this, because new aircraft can only be created over a development time of many years, manufacturers hope for production runs measured in decades. Following on, the development cost becomes a part of the unit cost for the client airline. So the lifespan for each unit that comes off the assembly line (including the last airplane produced of each model) is also a matter of decades. On page 39 of its *Global Market Forecast 2011–2030*, Airbus SAS confirms this:

> Due to the long term nature of the industry, the typical development cycle time for a new civil aircraft program takes ~ 10 years including time upfront for technology development, with high investments requiring design of product life targets of over 20 years, with a program life of about 40 years, this activity is focused on current products, the next generation of aircraft and even longer term, 2050 and beyond (Airbus SAS 2011).

If this seems incredible to you at first blush, think about some of the aircraft types in which you have perhaps flown: the Airbus 320 project was launched in the mid-1980s. That family of aircraft (A318, A319, A320 and A321) continues in production, and a design update, the A320neo (new engine option), is planned for release in 2015. This would mean that some kind of A320 aircraft would likely still be flying in 2045 or beyond—six decades. As for Boeing, the B-737 rolled out in 1967. Several versions are still in production (Boeing Company 2014a) and

---

4   Fewer than four scheduled departures per million are involved in an accident, and only a small fraction of those result in fatalities (International Civil Aviation Organization 2013a).

will presumably fly until at least 2040—longer still. The B-747, having started life at about the same time as the B-737 (mid-1960s) has been updated with a new version, the 747-8, just entering service in 2011 (Boeing Company 2014b). It is this kind of program longevity that warrants such huge expenditures in design and development. And in-service life measured in decades justifies the purchase price for the airline as well. Of course not every aircraft type is a jackpot winner in terms of its success in the market, and the risk and cost are reflected in the price.

With that in mind, it should come as no surprise that each individual aircraft represents a huge financial commitment. Of course the largest aircraft, with the most extensive tooling, sold in the smallest numbers, will reflect this reality most starkly, but all airplanes are expensive and a large component of the cost is the investment in creating the type. Airbus lists the price of their largest transport aircraft, an A380, at US$414.4 million. A smaller A318? Just US$71.9 million (Airbus SAS 2014).

One particular element of the security of this major investment, and an important factor for developers, is that some of the major technologies that will be applied have already been well amortized. Application of those technologies is then cheap compared to other development costs. The fuel system and engines are *major* core components of aircraft. Design elements of these components are adapted to handle and burn an extremely narrowly defined kerosene jet fuel. To introduce the use of a new fuel would require not just newly designed engines and systems, but designs that catered for new assumptions. Again, if we achieved commitment to that project, it would be many years before we saw the introduction of such aircraft and many more years—stretching to decades—before the fleet was built out in that new way. And modification of existing aircraft would, in some ways, be an even bigger job.

Then there are the logistic difficulties. Deployment of the first cohort of such new airplanes would need to be limited to a small route network where the costly and disruptive introduction of the necessary separate fuel storage and delivery system had been made available. The fleet and route network would then gradually be expanded as existing aircraft were modified and fleets were renewed to bring airplanes designed to burn the new fuel into service. And all of this upheaval would be done despite the fact that kerosene was adopted in the first place because it was found to be the ideal substance. That means the economics of the airline and aircraft manufacturing industries would be taken off optimum at a stroke.

It *is* doable. And if it were required, the aircraft and engine makers would figure out how to do it, and the airlines and their customers would pay for it. But it is challenging in the same way that a wholesale shift to something totally different would perhaps be—though obviously not likely to the same degree. So this research and change will not be undertaken if there is any other way of solving aviation's flight energy requirements and environmental and social obligations in a manner that makes more practical and economic sense. We simply cannot ignore the shorter term.

Is There Any Other Path?

'If only there were a way to make kerosene out of something else,' one might wonder. This thought frames the issue appropriately, because:

- if kerosene is an ideal fuel, and
- if aircraft and engine design have evolved so that only kerosene can be burned, and
- if it would be time-consuming and expensive to change aviation's dependence on kerosene, then we need to know if there is some way to change not the fuel itself but rather where and how we could get that fuel. It turns out that there is another way; in fact there are many. And getting to it is looking easier, quicker, and cheaper.

In the next chapter we will discuss how not just one but many technological pathways have been developed for such 'new-source' kerosene. So we do not need to waste many years to find a fuel material and convert the fleet to burn it, we have found many materials and many ways of converting them into just what the fleet is already designed and certified to use. I am underlining the facts and logic here partly because media reports have been misleading on the subject. Research in this area has led to headlines such as 'XYZ Airlines tests new biofuel: B-747 flies from A to B on jatropha oil.'[5] In fact, these are generally not tests of alternative fuels at all; in most cases the 'new' fuels are essentially identical to the petroleum-based kerosene that jets have always burned. Nonetheless, these are demonstrations of a much more impressive feat of science and technology: we have figured out how to make kerosene from almost anything.

---

5   For example, see International Civil Aviation Organization (2012a).

Chapter 4

Jet Fuel: One of Alternative Energy's Orphans

Around 2005 or 2006, when it started to become common knowledge that we could make hydrocarbons rather than mining them, everyone except the fuels science cognoscenti was startled. Though it is still a remarkable story that continues to surprise many, it is not the focus here; in the face of the larger policy, economic, and commercial issues that now arise, it is more important to properly examine how we will best benefit from what the alternative source hydrocarbon technologies seem to offer. Nevertheless, it is worth remarking on the significance of these discoveries. It will be impossible to understand how to react to and capitalize upon the developments in fuels science unless we have some grasp of their nature, maturity, and viability.

The Nuts and Bolts of Alternative Sources for Hydrocarbons

The notion that useful hydrocarbons come to us from under the ground as some distilled or processed fraction of petroleum oil, gas, or coal has a prominent place in our thinking about the economy and the commodities that supply it. It is therefore a powerful revelation that our important traditional fuels can be acquired in another way, particularly when attempts to transition to different fuels have not been entirely convincing. And it was this knowledge, as it became clearer over the last decade, that so moved the imaginations of people in the air industry.

A lot was going on. The movie *An Inconvenient Truth* (2006) was in theaters and television, and print media were full of the information that brought a public dawning of awareness and attention to the issue of global warming. The air travel industry, already feeling particularly vulnerable to the negative image implications of this new focus on GHG emissions, was targeted by people insisting that flying should be penalized and reduced. Into this mix came the news that we could get fuel in a way that was not so extravagantly productive of carbon dioxide (CO_2) and other pollutants. The notion was even mooted that truly 'carbon-neutral' fuel might some day be possible. On the other hand, many had been disappointed by the relatively modest environmental benefits realized by alternatives like corn ethanol. So folded arms, rolled eyes, and guffaws were in evidence—understandably. For some people, this is still the current situation. A certain level of cynicism toward novel liquid fuel sources is inevitable at this stage, and it is worthwhile to clarify what distinguishes one technological innovation from another and from past renewable fuel technologies.

Credibility is always critical, and let us hearken back to what we said earlier about technology: Credibility can be difficult to achieve when putting forward technological fixes to a global climate crisis that was itself brought about by unrestrained technological development and its attendant exploitation of immense resources of energy, including the fossil fuels that release the carbon with which we are now contending. Earlier efforts to get away from fossil fuels by using 'renewables' have created their own credibility problems. But the basis of the whole discussion is that renewables—in some form—*should* work. But what renewables? Produced how?

Just on that point, it is worth reviewing this one essential fact: all carbon-based fuels produce CO_2 when burned (oxidized). The difference is that fossil fuels release carbon that the Earth had managed to sequester (as the fossil remains of life) millions of years ago. In that sense, this is 'new' carbon *for us*; burning it, in the form of coal, or petroleum, or natural gas, adds to the current atmospheric CO_2 burden. But the fuels that will be discussed here are different. It is the goal that, ultimately, they can be made in ways that *remove* carbon from the atmosphere, so when CO_2 is released upon combustion it effectively recycles, leaving the atmospheric CO_2 level at its former level. Many past efforts to achieve this have been simplistic in conception, with poor anticipation of the energy requirements of the fuel production cycle: in some cases it took almost as much fossil fuel energy to make the new fuel as, in the end, would be produced. That is why many people in the environmentalist community hate hearing the words 'renewable fuel' or 'biofuel'. Fair enough.

As the aviation industry is telling itself, and admitting to everyone, air travel will be demonized to the point of significant loss of revenue unless it can show that it is using fuel that is significantly more sustainable than both petroleum-derived kerosene and other fuel sources that were (falsely) claimed to provide huge environmental benefit, and did not.

So, in the context of the public's somewhat cynical attitudes toward renewable fuels, and fears about technology's possible unwillingness to clean up after itself, it is important for everyone in the transport sector as well as researchers, students, and policy-makers to actually have some technical grasp of what fuel strategies are under consideration, how they are intended to operate, and how they differ from past endeavors like, for example, corn ethanol.

A Note on Organic Compounds

We will see that there is more than one way to synthesize jet fuel from non-fossil feedstocks, but it would be silly to ignore the fact that a large number of these ways do fall into the category of 'biofuels'. Inasmuch as a great deal of this chapter will be about biological sources of raw material for fuel, it is worth repeating that the hydrocarbons we have used until now have been sourced from the remains of plants and animals that lived on the planet's surface millions of years ago and have been undergoing subterranean heat and pressure treatment ever since. So perhaps it should not come as a surprise that scientists have figured out a way to eliminate the long wait, and to successfully process material of biological origin (biomass) in a manner that strips it of all but its essential carbon and hydrogen. This is, in effect, taking biomass currently available at the Earth's surface and accomplishing quickly in a processing facility some of the same transformations as those produced over millenia underground. The growth of plants draws down atmospheric stocks of CO_2. So making fuel from plant biomass allows us to exploit existing atmospheric carbon and leave unused the carbon that has been safely stored below the Earth's surface as petroleum, oil, or gas.

This particular strategy underlines an essential bit of knowledge, which is that virtually all current biofuels science involves manipulation of organic compounds—molecules that are made up of carbon atoms (C) together with other elements such as hydrogen (H_2), oxygen (O_2), nitrogen (N), sulfur (S), and other substances as well. Jet fuel demands the specific type of organic compound mentioned above: hydrocarbons. Hydrocarbons are molecules in the forms of chains or rings of carbon and hydrogen. We can obtain hydrocarbons where they naturally occur (as we do now), synthesize them from scratch using component atoms of carbon and hydrogen, or create the correct assemblage of hydrocarbon molecules by manipulating other organic compounds so that the result is the specific hydrocarbon that we seek.

If mining these compounds is no longer an option because it causes the release of long-sequestered carbon into the atmosphere, we are left with the other possibilities: either create them directly or modify the molecules of substances that already contain them. The direct creation of hydrocarbons from their component carbon and hydrogen atoms is probably the most challenging path, but even that is now receiving some attention. If we are able to do this in an efficient and economically viable manner (and efforts are under way), we will be making purely synthetic fuel, not biofuel.

However, much of the current research into the third strategy (modification) resembles the adage 'How does one carve a statue of an elephant? Simply take a block of stone and chip away everything that doesn't look like elephant!' This is the approach being taken toward the kinds of biofuels that aviation might use—take other organic molecular structures and remove the elements that keep them from being jet fuel hydrocarbons.

What's In A Name?

Before I go further, I want to outline a couple of extremely important points related to vocabulary. Since the 1980s the term 'biofuels' has come into widespread use, and for better or worse, will continue to be used as the word that describes so much of what will be undertaken in this field, although its definition is going to be stretched. First, biofuels are popularly understood to be either substances that can be used directly as fuel (vegetable oil burned in road vehicles would be an example) or substances created *from* biomass using well-understood biological and physical process technology pathways (corn sugar undergoes biological fermentation and distillation to become alcohol, for instance). Significantly, in both examples, the resulting fuels are different from the ones that they are intended to replace.

With that, the definition should probably have hit its limits. And neither type of fuel production would have much application in jet fuel (with at least one notable exception; of which, more later) .However, 'biofuels' is in full use in jet fuel terminology and includes substances such as hydrocarbons created by stripping oxygen out of plant triglycerides, or hydrocarbons fermented directly from sugars or plant cellulose using new microbes, or hydrocarbons created by turning all components of a given biomass (fiber, sugar, starch, oil—the lot) into carbon monoxide (CO) and hydrogen 'syngas' which is then chemically reassembled as a hydrocarbon. These are a few examples only, but in all such cases the resulting fuel can be exactly the same as the one it is replacing. So, in the case of jet fuel, the new fuel is the same as the old, but the way of acquiring it is entirely different. We might wind up calling it biofuel or biojet but it is the same fuel, and 'bio' means a host of different things. I support the approach of referring to such fuel as 'sustainable jet fuel—differently sourced', or 'sustainable alternative jet fuel', or words to that effect; they are 'drop-in' fuels.

With regard to the vocabulary that is evolving, I do not use such qualifiers as 'second generation' (or 'first', 'third', 'fourth', or 'next') biofuels. Others have no problem with such terms, but it is my humble view that this generation-numbering business has no agreed-upon definition and is making discourse more confusing rather than less. It is not terminology that is used in a consistent way. Within individual groups, such terms may have agreed-upon meanings, but only some people are familiar with them, thus discouraging broader understanding among larger communities.

Another important point, mentioned a bit earlier, warrants expansion: Not all novel sources for jet fuel will have a 'bio' component in the feedstock. Current research and development paths include techniques such as using chemical energy and materials inputs where the only 'bio' element is microbial agency, and the fuel is assembled from carbon and hydrogen molecules. In other words, there is a microbial (biological) mechanism that processes raw material into fuel but the raw material is not necessarily biomass.

Here is another well-worn term: 'alternative fuel'. For many people, this would normally mean alternative fossil feedstocks for liquid hydrocarbons, perhaps gasoline refined from coal or natural gas. But now 'alternative' may be applied to the results of a host of processes that have nothing to do with fossil fuels.

Needless to say, this lexicographical mix-and-match quagmire results in a lot of miscommunication, exacerbated by regional preferences that favor one definition or usage for a term over another. For this book, I will try to identify the various aviation fuels on the basis of the actual materials and processes used to create them, but it will sometimes be expedient to lapse into use of 'biofuel' and 'alternative fuel'. Forgive me. In any event, it is time to start discussing them. And since most of the increased-sustainability fuel options under discussion could be called some form of biofuel let us start with those.

It is hard to keep up with developments in this subject area. Everything that this book discusses has changed since its writing was commenced. That is particularly true of fuel manufacture and feedstock production technologies. Virtually all of the value of the next sections relates to the fact that there are many production options, rather than to what those options might actually be. So the following summary is neither exhaustive nor current in any real sense, but like anything else, reading about these things helps us understand others.

Biofuel Technologies

Fischer-Tropsch (F–T), 'Biomass-to-Liquid' (BTL)

We start with Fischer-Tropsch technology because it has a long history in alternative fuel production and because the F–T process is an exercise in deconstructing and reassembling organic molecules in a very fundamental way.[1] Talking about F–T acts as a kind of primer on the larger subject.

F–T fuel processes are actually comprised of two distinct stages: gasification (the production of a synthesis gas or 'syngas' mixture of carbon monoxide and hydrogen through the application of heat and controlled exposure of raw material to water or oxygen), followed by F–T synthesis of the desired materials or their precursors.

All of this technology is quite old. Gasification was already well known by the 1800s, when it was being used to create syngas ('coal gas') for municipal lighting purposes. But our interest is not in the direct use of such gas as fuel; in this context we are talking about using the gasification product syngas as raw material.

Franz Fischer and Hans Tropsch started their work in the 1920s, and their process of using syngas in the creation of other substances was one of the technologies used to synthesize aviation and road transport fuels where supplies of conventional petroleum were disrupted due to war or embargo. Some of Germany's ersatz fuels during the Second World War were produced in this way. South Africa made great strides with the technology during the apartheid embargos.

But in those cases the problem was security of supply of petroleum, not carbon emissions, so the raw material was another hydrocarbon that happened to be more plentiful—coal or natural gas. In the biofuel context, renewable biomass is the raw material that is gasified to produce the required syngas. The basic carbon

---

1   For further information about Fischer-Tropsch history and technology, see Stranges (2014), http://www.fischer-tropsch.org

monoxide and hydrogen components of syngas include the requisite hydrogen and carbon that might be used to make a hydrocarbon. We just need to know a little more about how they become part of syngas and how they can then be re-combined to produce a fuel.

When we look at the chemical formulae or structures of all common plant molecules, we notice three elements: hydrogen, carbon, and oxygen. Sugar and starch (carbohydrates), cellulose and lignin (plant fibers), and vegetable fats and oils (lipids) all differ from our desired fuel hydrocarbons in that they contain oxygen. Plants do contain many elements or compounds that we need, but the major hurdle is that the molecules of interest—the ones that make up the bulk of the plant's mass—include oxygen content that renders them less useful as fuel. In the case of airplanes, it makes them completely useless.

Hu, Yu and Lu (2012) have provided a useful summary of current biomass-to-liquid technologies using F–T synthesis. In outline, to produce the syngas required for F–T synthesis of target fuels or precursor substances, the biomass gasification process must be tailored to the characteristics of the biomass—for instance, the moisture content and particle size must be accounted for. Gasification always involves raising the temperature of the biomass and exposing it to steam or O_2, sometimes in the presence of a catalyst. The products of gasification are typically varying proportions of CO, CO_2, H_2, a small amount of methane (CH_4), and various impurities. The CO_2 and impurities are removed in a gas cleaning process. The remaining CO and H_2 are combined in the F–T synthesis reaction to produce hydrocarbons and oxygen, and every process that turns plant biomass into hydrocarbons for jet fuel must include a further step to remove that oxygen.

Since the waste CO_2 from the gasification process and the CO_2 exhaust from the ultimate combustion of the fuel in the airplane both came from a (presumably) renewable and sustainably produced biomass which continues to be cultivated and can reabsorb that carbon again, no net CO_2 is produced unless fossil fuels are used as an energy source in some part of the process supply chain or mechanism. I will allow right here that we are not on the point of being able to produce even carbon-neutral fuel because of the way that we power such processes. But that is changing. Another twist: Some sustainable fuel technologies require a concentrated stream of CO_2 and perhaps biomass gasification could provide that source.

I want to emphasize an earlier distinction. Some F–T fuels were strategically and economically important in the context of their development from fossil fuel feedstock in the mid-twentieth century. We must remember not to confuse F–T biomass-derived BTL fuels described above with those that rely on hydrocarbon raw material like coal or natural gas. Unless additional carbon capture and storage (CCS) technologies are incorporated, 'Coal-to-liquid' (CTL) or 'gas-to-liquid' (GTL) fuels have an even higher carbon impact than current conventional petroleum distillates (Reed 2007). And in any case the CTL or GTL process demands energy that is often, itself, carbon-intensive. So F–T fuels of this type are really only useful in addressing security of supply or stability of price issues; they do nothing for the global warming problem.

Hydro-Treated Renewable Jet (HRJ)

The oils and fats and some other substances in plants (generally referred to as lipids) are a potent source of energy. Triglycerides are an example. Their chemical composition makes them very useful: the 'tri' part indicates that they are essentially made up of three long-chain fatty acid hydrocarbons. The chains are tied together at one end by a more complex part of the molecule, making the compound's structural diagram look rather like a very small hand with just three long fingers. The glyceride (palm) part of the structure is essentially an incomplete glycerin molecule (or glycerol), which lacks three hydrogen atoms. If we supply those hydrogen atoms (plus an elevated temperature and a catalyst), the glycerin molecule is completed and casts itself off, releasing the three dangling fingers of hydrocarbon molecules (and any other by-products) at a stroke.

Figure 4.1 allows us to see the transformation that occurs during the transformation of plant oil. But it should also serve as a general illustration of how hydrocarbons can exist 'within' molecules of biomass.

This is a good time to mention that in all of these pathways to our desired hydrocarbons, the hydrocarbons that we initially obtain are not always the right size or shape, and must sometimes be 'cracked' or broken into smaller carbon chains by the insertion of additional hydrogen ('hydro-cracking'), or alternatively they need to be polymerized, whereby smaller molecules are linked together to form larger ones.

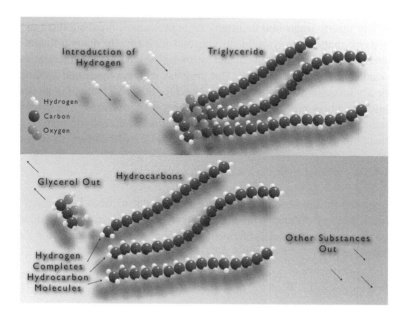

Figure 4.1   Example of hydrocarbons 'within' plant oil molecule

HRJ fuels are made only from fats and oils; naturally occurring oils must first be mechanically or chemically released from the biomass, which can present quite a challenge. For example, one biomass source very high in oil content is micro-algae, but it is extremely difficult to extract the oil from these tiny plants. That is changing, however.

Sugar Fermentation, Distillation, and Catalytic Conversion of Resulting Ethanol

Until recently sugar was thought to offer little in the way of jet fuel possibilities. It can be converted into jet fuel through gasification and F–T process just like other biomass sources, but its other pathways to fuel have been less apparent. However, there are a couple of options, and one is a process whereby ethanol is passed over a special metal-impregnated zeolite catalyst to become a collection of fuel precursor hydrocarbons. The initial ethanol is derived from yeast fermentation of sugar: the yeast metabolizes the sugar in a water solution and excretes CO_2 and ethanol, which is then purified by distillation.

Fiber Cellulolysis

While some sugars may be difficult to break down, plant fibers (lignin and cellulose) are undeniably even trickier. We know that ruminants like cows can digest fibrous plant materials and convert them to nutrient sugars. The fibers are first weakened by long chewing, then bacteria in ruminants' digestive tracts secrete enzymes that cause the fibers to disintegrate further through a process that chemically inserts water into their very large molecules, breaking them up. This is called enzymatic hydrolysis. Identifying and analyzing the many different types of bacteria and enzymes involved, and synthesizing them for commercial use has been challenging, to say the least. But progress is being made and a number of proprietary technologies have been developed for the processing of plant fiber.

Once the fibers are broken down and the sugars obtained, it is possible to proceed to the fermentation to ethanol, distillation, and catalytic conversion of the ethanol to hydrocarbons, and conversion of those hydrocarbons to kerosene.

Microbial Combined Hydrolysis and Fermentation

As suggested by the heading, this process is similar to what has already been described. The key difference (and it is a big one) is that microbes have now been created that will hydrolyze *and* ferment plant fiber so that it is converted directly to ethanol.

Pyrolysis

Pyrolysis is a heat-based process that is applied to bulk biomass: oils, sugars, and fiber; it is most useful in targeting mixed biomass that has a high fiber content—

wood waste and the like. If we were to burn such a substance, the ongoing heat of combustion would cause the more volatile substances to migrate out of the fibrous structure, where they would then combine with ambient oxygen and ignite. The heat breaks down the rest of the material. But if wood is heated in the absence of oxygen, ignition never starts and the escaping volatile substances can be condensed and captured as 'pyrolysis oil'. The pure carbon that remains as residue (biochar) can be either used to help power the process or sequestered as an agricultural soil supplement, for example. Pyrolysis oil (which is actually a mixture of many organic compounds and water) can, in part, be refined and reformed into useful hydrocarbons including the target fuels. In some pyrolysis operations, by-products can also be gasified for F–T fuel processing.

Direct Microbial Fermentation of Sugar, Fiber, and Protein to Kerosene

Recently, it has become possible to create bacteria that can consume sugar and convert it directly to kerosene (Steen et al. 2010). The lighter-density kerosene floats out of the aqueous sugar solution, thus allowing the microbe habitat to remain livable and permitting a continuous conversion process. But new tailored microbes are being developed that can do the same thing with fiber and plant protein. This may prove to be an excellent technological biomass pathway to fuel on a standalone basis, but is also interesting in terms of converting by-products of other processes. For example, the economics of algal biofuels might be greatly improved if fiber and protein waste could be processed into fuel after the valuable vegetable oil was itself extracted and converted to fuel.

Comment and Disclaimer

The foregoing sample is neither comprehensive nor detailed, and cannot reflect the rapid pace of development. But if we recall that very little about any of this was common knowledge just a few years ago, we get some idea of how quickly things are moving. It is extremely difficult to know, from one moment to the next, what combinations of biomass and technology may represent the most significant progress toward more sustainable fuel. The rapid addition of new elements to this complex mosaic makes a snapshot not only difficult but also rather limited in value.

Figure 4.2 is intended to reflect the complex nature of the dynamic rather than its current state. Certainly many of the technologies described in this chapter are not ready for immediate scale-up and commercial deployment. In fact some are merely ideas that have proved out in a laboratory but nowhere else. Taken together, they represent some of what is being talked about in biofuel science today.

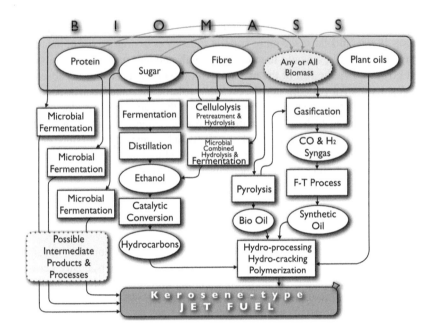

Figure 4.2   Some process pathways from biomass to jet fuel

The rate at which new ideas are being proposed or essayed in an energy-hungry world is really quite astonishing. And while we could perhaps dig harder for more of the biofuel options that are showing some promise, it serves little purpose: this is a boiling pot and there is just no point in trying to describe all the bubbles of ideas that have made it, or might make it, to the surface.

Stretching the Meaning of Biofuel

To draw a line under the previous point (and to elaborate on an earlier passing reference) it would be useful to mention a technology that takes us beyond what we can reasonably call biofuel science. As an example, one commercial enterprise has engineered microbes that can convert carbon dioxide and water into specific fuels without exploiting any biomass at all (Joule Unlimited 2014). This could be called a biofuel inasmuch as it is a living organism that produces the fuel, but the process is not fed by biological feedstock. The only other input would be light energy, constituting a synthetic shortcut for what might otherwise be a combination of plant photosynthesis plus biomass harvesting and processing. Another technology that uses waste flue gas and proprietary microbes is at commercial demonstration scale and is described in a later chapter. All of this takes us into a different realm: while biological elements do play the key role in the actual operation of the process, there is no involvement of biomass as a raw material and so all of the sustainability concerns that we normally associate with life-cycle analysis for carbon in the feedstock are not pertinent. We can say that the biological element would be part of the process but not part of the fuel.

Speculation about where such technologies might take us cannot be seen as frivolous; imagining what we might be able to do in a commercially viable way over the coming decades will become a vital, continuing part of the work. If we consider that some fuel production ideas exploit flue gas, for example, future versions of this or other processes might be able to use ambient atmospheric CO_2. Then the greenhouse gas situation could change a great deal. This is admittedly a stretch, but given the previously discussed dire circumstance that we find ourselves in concerning atmospheric CO_2, we should keep a fairly close eye on the prospects for carbon-neutral and perhaps even carbon-negative fuels. The point, and we will have lots of discussion about its relevance in chapters on policy, is that we are in this for the long haul. What seems likely, based on what we have seen of the evolution of alternative fuels technology in the recent past few years, is that we have *very* little idea about what we will be using to make fuel a few decades from now.

Carrying on, we have to think about such things as the exploitation of synthetic biology. We have just talked about the use of proprietary microbes in converting CO_2, but now under discussion is the notion of so-called 'electrofuels'. The US Department of Energy's Advanced Research Projects Agency–Energy's Electrofuels (ARPA-E) program was established in 2010 to explore the possibilities of 'using microorganisms to create liquid transportation fuels in a new and different way that could be up to 10 times more energy efficient than current biofuel production methods' (United States Department of Energy. Advanced Research Projects Agency–Energy (ARPA-E) 2014). The basic plan is to create liquid fuels from CO_2 by using energy extracted from fairly mundane materials.[2]

This is probably as thorough a summary of technology processes as we need for the purposes of this book (I can hear many saying it's more than they ever wanted to know). We are beyond conventional fossil fuels, unconventional fossil fuels, biofuels, or even a stretched definition of biofuels.

But before we move on to the next topic we should discuss not only the fact that there are varying technologies, but also what materials might be needed to feed such production methods: feedstocks. Processes that use common, abundant materials that can be sourced without environmental or social implication are ideal. But since most new alternative fuel ideas are biofuels, and inasmuch as most of the processes described rely on a biomass feedstock, let's come back to them, carry on to the next step, and get an idea of some of the potential sources for biomass. We do this for one overwhelmingly important reason: Sourcing and transporting biomass feedstock is, by far, the biggest cost area for alternative fuel producers.

---

2   A more detailed description of the electrofuels project is provided by Conrado et al. (2013).

Elaboration of Biomass Options

Coming back to biofuels specifically, keep in mind that some processes accept any kind of biomass: virtually any organic material is a potential feedstock. But let's look at some particular categories and sources of material.

Waste

Knowing that sometimes anything 'biological' can be used, one of the first and most important things to consider is waste. Just about all kinds of organic waste matter can be turned into fuel in a laboratory. It is not perhaps so straightforward in the real world, but even here waste can be a viable feedstock. Waste streams can be dedicated to provide a homogeneous type of biomass or other organic feedstock, and processes can be tuned to that type of material. This is the case with wood waste, sewage, plastics and a host of other materials. Technology can be improved to accept a wider range of feedstocks—gains of that sort would be needed in order to process unsorted garbage streams. Fiber-consuming processes are good candidates for many types of waste. We produce waste in all aspects of our human activity and so the potential sources are limitless. The one thing that we should remember is that waste that would otherwise be processed in a way that permanently sequestered its carbon content would not be a good choice for making fuel which, when burned, *releases* that carbon. If a piece of paper or wood were just going to be left to rot in the open, it would be preferable to turn it into useful fuel. On the other hand, if it were going to be sealed in an airtight environment forever, from the atmosphere's point of view, that would be better. Since feedstock cost and transportation form the largest part of alternative fuel production, waste is wonderful in that it is usually located near to centers of human habitation and low in price.

Oils and Sugars

Plant oils are a very practical source of fuel material from a process point of view. A cursory read of popular media or biofuels literature will include reference to oils from plants like brassica carinata, castor, camelina sativa, canola, coconut, jatropha, jojoba, micro-algae, palm, rape seed, safflower, soybean, and sunflower (listed alphabetically) as potentially commercially viable sources of plant oils but this is not, by any means, an exhaustive list and is intended only to show that the options are vast. Sources of sugar with which we have become familiar include corn, fodder beet, Jerusalem artichoke, sugar beet, and sugarcane.

Fiber

One of the most interesting potential sources of biomass, only because it is often less likely to be used in other aspects of our economy but is produced in large quantities nonetheless, is fiber. Fiber can be purpose-cultivated, although plentiful amounts of unexploited plant fiber remain after cultivation of plants for protein, oil, sugar, or starch. It is not free, however; replacing the soil nutrients removed with the waste fiber can represent a substantial cost. Bagasse, canary grass, corn stover, elephant grass, hemp, miscanthus, straw, switchgrass, and wood all constitute potentially valuable sources of plant fiber.

Scoping the Concepts

What does it mean to have so many ideas before us? How do we deal with various fuel technologies and how do we keep track of what they offer us now or what they might offer at some future date? We should take a moment to appreciate not only the wealth and variety of available ideas, but also the policy challenges that such a wealth of ideas may involve. It may seem daunting and paralyzing to be confronted with so many choices, because commitments to one technology over another could prove costly. Nevertheless, we should see all of this as an opportunity. We also need to remember that policy must be crafted in a way that encourages the development of the very best ideas on offer and avoids ongoing commitment to lesser fuel technologies that may enjoy political support from groups with particular vested interests.

In that regard, remember also that the fuels technologies described vary from the mature to the tentative, are responsive to differing feedstocks and energy sources, and will come online at different times, offer different advantages, and be amenable to some very particular climates, economies and labor resources. So while not every idea will come into useful development, many might, and it may not be possible or desirable to choose just one or two ideal candidates from among them. We may very well have to adapt ourselves to our options as our options adapt themselves to our circumstances in time, place, resources, and wealth. We will discuss this in much more length in the chapters that touch on policy.

A small discussion of potential biomass sources allows us to relate them to some of the technologies that have been described. But the categories—oil, fiber, and sugar—are admittedly simplistic; the materials' compositions are a little more complicated than that, and there is also a great deal of overlap: plant fiber will almost always be found in addition to whatever other material is considered prime.

Never Mind 'Bio': Fuel From Air and Water (!?)

We have talked about technologies where the 'bio' component is limited to process technology rather than feedstock, and how that really does take us farther than what we normally call biofuels. To emphasize the point about our information being a snapshot with a wide-angle lens, ideas for flight energy go beyond almost anything that the uninitiated would contemplate or even dream about. And they certainly go beyond anything with even a whiff of bio.

For one: as has been mentioned, it is theoretically possible to scrub CO_2 from the air and combine it with H_2 acquired through the electrolysis of common water. Renewable solar and wind energy can be used to power such a process. This would turn a problematic, variable, intermittent source of electrical energy into a relatively steady source of liquid chemical energy—a truly alternative flight energy source. Or this: the Royal Society's 2011 Summer Science Exhibition featured a presentation of nanotechnology being used to create solar fuel or fuel component elements (Royal Society of London 2011).

This is why we want to prevent the conversation from being framed in a way that implies that biofuels are the only answer or perhaps even the best option, in the long term. The best option is always the one that offers improved atmospheric carbon reduction together with reduction in other general wear and tear on the global and local environments, fair treatment of people who need work and, of course, produces fuel at the lowest possible cost!

The foregoing chapters constitute, perhaps, some of the raw material of the larger discussion: the magnitude of the industry's emissions challenge and the things that can be done to reduce emissions. We touched on the scope for improving aircraft, operations, and airspace management. We examined the limitations on bringing a wholesale alteration to propulsive technology, and so why we must continue with kerosene. But we also learned that we can acquire it in a new way. The next bit of the preliminary part of the book describes the policy context that will constitute the setting for everything that we must do to achieve the goal of sustainable flight energy.

Chapter 5
Policy: Background

By one kind of logic, this chapter is a bit out of place. We are a good way through the book and we have not really dealt with one of the title subject words: Sustainability. Yet everything that we have examined thus far implies that it is important. We will do that. It will occupy an enormous part of the discussion. But for two reasons it was considered desirable at this point in the book to talk about where policy has been. The first reason is that it allows us to make a break between the history and background aspects on the one hand, and everything that must come next on the other. The second is that it scratches an itch: we know enough now about the fuel problem to wonder what has been done to advance solutions.

The fact is that *some* policy already exists now. Having seen the size of the most obvious aviation sustainability problem, why we are in it, and a few of the things that might help us get out, this chapter will let us understand the other part of the background: policies that already affect aviation and GHG emissions.

In Chapter 6, we will embark upon understanding sustainability as an idea, and then move on to talking about ways of assessing measures of sustainability and ultimately how to make it happen. But let us complete the essential backdrop.

Emissions Policy History

With a few exceptions, we do not really have much in the way of policies that speak about aviation's carbon profile. What we do have are policies that frame both the way the industry operates and how the larger global non-aviation-specific discussion about emissions has proceeded. The policy settings concern past international negotiations (and their results) as regards:

- commercial access to air routes, and
- how jurisdictions (including aviation) should address their obligation to reduce emissions.

The former doesn't immediately seem to be important to our topic, but it is indeed relevant. The connection is this: aviation's treatment on the commercial policy front allows us to detect where 'levers' may exist, and where they do not. Demanding (forcing, coercing) improved levels of sustainability for aviation and ensuring that solutions can, themselves, be created in a sustainable way requires understanding how the industry addresses commercial realities worldwide. The development and operation of mechanisms for controlling commercial access to countries and their airports is at the very center of this matter. The reason is that policies have economic and financial implications for all of the concerned parties, and the commercial relationship between each airline/host government pair and all of the others is a matter of profound consequence. International or bilateral agreements are reached in order to allow governments (acting in favor of their own carriers) to pursue their interests, and some of the most important interests are commercial. Bluntly, what we can require of each other is based on how (positively or negatively) we can affect each other.

Domestic Versus International

We will learn a lot more about the commercial part of this subject, but there are a few preliminary details that we should know about the emissions policy part. First, aviation carbon emissions are not all on the same footing. The carbon emissions attributed to domestic air travel form part of each country's respective national emissions budget. Within a given country, nothing makes domestic air travel different from any other GHG-generating activity such as road transport, for example. Discussions among countries about their respective emissions budgets and the actions that they will each take to reduce national emissions are infamously problematic: some countries are interested in dealing with the problem, some are not; some have started meaningful efforts to create national policies that will bring emissions down, some have not. Domestic air travel emissions are matters for the national debates about their share in reducing global emissions. We can imagine that anything that comes along that will help make flying more sustainable will apply wherever flying is done—both internationally and domestically. But that will happen on the basis of dozens of domestic, national, respective policy regimes.

A critical piece of this policy puzzle is that *international* air travel emissions (along with international maritime transport emissions) have *not* been lumped in with national emissions budgets in discussions thus far. GHGs from international air travel have, to date, been considered cumulatively as a completely separate body of emissions, to be managed as though 'international' were a country on its own. So the international arena is where policy might have been brought to bear specifically on air travel. (The reasons for this separation will soon become clearer, when discussed in the context of the Chicago Convention and the commercial component of the policy issue, so it will be a recurring subject in various contexts as the story evolves.) I would say that everything related to aviation and sustainability is important. But from a policy development perspective, this domestic/international consideration is absolutely key.

Another thing to keep in mind is that when we speak about policy as it has evolved so far—particularly at the international level—we are talking mostly about who is responsible for carbon and how they will be held to account. Generally, we are *not* talking about any kind of policy that would assist with reaching the goal of actual lower-carbon operations. We will see later on how this distinction becomes relevant. It is true that aviation emissions solutions (no matter where or how they develop) may very quickly become available at all scales and for all airline operations, domestic and international, in all countries. But with domestic air travel emissions considered to be part of a country's national profile, and each country having a different attitude toward its respective total national carbon reduction responsibilities, some of the most fertile ground for creating and implementing an aviation-specific way of reducing emissions, and the policies that would support it, are found at the sectorized international scale.

Note another fundamental: when we say 'international aviation', we are talking about the emissions that result from the actual activity of air transportation—flight operations and fuel use. The 'land-based' emissions produced during an aircraft's manufacture or the provision of infrastructure and other supporting services and materials form part of the domestic emissions budget of the country where the emissions are generated.

Since the only context in which air travel is considered as a separate entity and where the aviation industry itself (rather than a national government) is held responsible for its GHGs is in international air operations, that is where broader industry effort has been focused and where comprehensive policies can arise. So everyone participates in the discussions. Aircraft manufacturers and ancillary service providers are vulnerable when penalties are imposed on operations, so they are keen to find solutions. And the focus of the search for solutions is also in this international arena, with awareness that technology and policy tools that can reduce air emissions at the international scale may also be workable at the national level. By the way, the reverse is certainly also true, but many countries have apparently not realized that solutions developed internally could be sold into the international segment and into other countries' domestic air travel systems.

The Reference Points for the Intersection of Aviation and Emissions Policy

Two agreements (previously mentioned) have already framed the policy discussion to some extent: the *Convention on International Civil Aviation*, better known as the Chicago Convention, or just 'Chicago' in the following paragraphs (International Civil Aviation Organization 1944a), and the *Kyoto Protocol to the United Nations Framework Convention on Climate Change,* or 'Kyoto', (United Nations 1998), are our anchors. The former embodied the pursuit of uniform high standards, equality and cooperation within the aviation sector, set out the degree to which aviation could be regulated at the international level, and also led to the founding of the International Civil Aviation Organization (ICAO); the latter established many of the guidelines to be used in international climate discourse, and, famously, addressed emissions limitations internationally. Chicago created an understanding of how the global air industry would function in a regulatory way; Kyoto defined how countries' carbon obligations would be regarded and set, how accountability would be allocated (along both domestic and international lines), that the 'Annex

I' (developed) countries[1] should take responsibility for finding ways of reducing air sector emissions generally, and specified that ICAO should be the agency through which the efforts aimed at emissions reductions in international aviation should be coordinated. We will discuss both of these agreements in more detail, but I want to point to the salient provisions of each document now as we launch into discussions of policy, because a poor understanding of why things went the way that they did in Chicago and Kyoto is bedeviling international aviation climate discussions even as I write this.

The first question goes to Kyoto's relevance. In the minds of many, Kyoto is a failed agreement. The United States (US) refused to ratify. Canada and Australia abandoned their Kyoto commitments. That is neither here nor there. We do not discuss Kyoto because it is determinative and dispositive of all issues, or even *any* issues. We are interested in it as the only document that illustrates how the global community started to flesh out international aviation's place in the emissions picture according to the context set by antecedent agreements. No country—whether it signed Kyoto or not—contests the way in which international aviation is considered separately from domestic producers of GHG.

The treatment of domestic and international air industry emissions as set out in Kyoto reflects the realities that had previously informed Chicago. Kyoto elaborated two preexisting relevant principles. The first has nothing specific to do with aviation, but rather responds to the historical responsibilities and current economic capacities of different states: Restrictions on states' total national rates of emissions (which would obviously include those from *domestic* air operations) must reflect their respective degrees of national economic development. All nations would have responsibilities, but they could not be exactly the same. The principle of Common But Differentiated Responsibilities (CBDR) for reducing emissions means that we all have to get our total emissions down, but some countries (wealthy, developed) must do more and do it more quickly. This is not illogical—regardless of current *rates* of emissions, developed countries are responsible for most of the excess CO_2 that is actually *in* the atmosphere.

The second principle, concerning the segregation of international emissions, is the one that we have previously mentioned. Of course this concerns aviation, and seems to align with the non-preferential standpoint of Chicago. In a little more detail and context: International air operations would be addressed on a sectoral basis; air travel emissions on international routes would not be attributed to an air carrier's country of base, origin, or destination.

Beyond the complication of trying to bring responsibility for international air and sea transport emissions into the fray, this provision responded to the fact that (and here is our point of intersection) agreement could not be reached in Chicago on a globally uniform way of organizing the commercial provisions of international flight. The implication of this 'weakness' of Chicago means that

---

1   Annex 1 countries are listed at http://unfccc.int/parties_and_observers/parties/annex_i/items/2774.php

rights to fly between countries are negotiated between pairs of respective countries. Any global agreement that had financial implications would distort the effect of those bilateral agreements. Further, Chicago submits that air operations must be regulated in a non-discriminatory and undifferentiated manner, as is made clear in its Preamble and in Articles 37 and 38 (International Civil Aviation Organization 1944a).

These general principles, as regards the obligations of states on the one hand, and international air operations on the other, have been upheld in all subsequent climate talks. While they are not inviolable, it might now be difficult to get agreement on a new paradigm. That is because, as we can see, the second principle of separating international aviation emissions from the various national carbon budgets actually does seem to flow *from* the non-discriminatory provisions of Chicago: If we are to treat countries differently and we must treat all of their international air operations the same, we need a distinct agreement for international aviation.

But there is a problem here: many developing countries regard airlines as a foundational part of their economies, and want emissions from their airlines' international air travel to and from their cities to be treated in a way that reflects their (Kyoto-described) CBDR status and rights. Speaking of international air emissions and differentiated responsibilities in the same breath somewhat conflates issues and begs the question, of course. Every party that accepts the CBDR principle knows that it applies to a country's domestic emissions (including domestic air travel emissions) and that international air travel emissions are excluded *because* they could not be attributed to a certain state. But that does not mean that countries will not try to construct another way of looking at the whole issue. We will see a little more clearly why that is when we get to our discussion about the progress (or lack thereof) in current international aviation policy discussions.

It is useful to look a little more closely at Chicago and Kyoto in terms of understanding the context and background of the talks as much as for the importance of their outcomes and their influence on subsequent policy discussions, such as the fairly recent and very important decision of the European Union (EU) to make all air operations, within and into the EU, subject to the EU Emissions Trading System (EU ETS). If we do not understand the reasons for the paths taken by Chicago negotiations on the one hand, and climate talks on the other, we will be poorly prepared to sort things out. We will not know which principles are amenable to change and which are not.

The Chicago Convention

The point that we should especially note about Chicago is that it did not, and really *could* not, have established any capacity for dictating the considerations that inform global commercial air rights. Nor, as a consequence, could it have much to say about pacts that are drawn up between any two particular countries. Obviously, the same considerations tend to foreclose even more tightly on agreements that try to go beyond bilateral contracts and into the global scope *now*. Emphatically, it is useful to keep in mind the matter of why that is. Current attempts to negotiate a global international aviation emissions regime bump up against the same barriers that confronted the negotiators those many decades ago when they tried to establish a commercial regime. Anything that has a larger context and effect in terms of commercial, financial, or economic components is hard to insert into a global air agreement because the money interests of all of the parties are quite particular. Why did it turn out this way?

Chicago: the Setting[2]

By late 1944, the Second World War (WWII) was coming to a close. Within the next year, Roosevelt, Churchill, and Stalin would meet at Yalta to discuss the post-war future of Europe and the world, and German and Japanese forces would surrender. In this atmosphere of transition and recognition of the inevitable need to rebuild international institutions, change and fresh starts were the order of the day. The war had accelerated technological development in many important respects. Managing new capacities (such as modern, efficient, international air travel) was seen as having enormous commercial importance, and would be a key element in an essential worldwide economic reconnection. Since all countries wanted to participate in this economic development and influence, an aviation conference was organized on short notice; in November and December, 1944, delegates from over 50 countries met in Chicago, Illinois to discuss both technical standards and commercial route and destination rights.

The first (technical) theme for the Chicago discussions related to the myriad topics associated with making international commercial aviation safe and (to the extent possible) consistent in such matters as navigation standards, the design and use of navigation aids and air routes, air traffic services, the provision of meteorological information, aircraft certification, pilot licensing, and everything else that warranted the attention of national regulatory bodies and that could usefully be standardized if international agreement were to be achieved. Building on the groundwork laid in previous conventions, Chicago produced a valuable result in this regard.

The second branch of discussion had to do with commercial aspects of international air travel, such as landing and overflight rights. In the period between

---

2   In providing details about this setting, the civil aviation history that led up to it, the proceedings of the Chicago conference, and the subsequent evolution of civil aviation policy and ICAO, a special series of twelve *ICAO Bulletin* articles by Duane W. Freer, the onetime director of ICAO's Air Navigation Bureau, is comprehensive, concise, and informative. Much of what follows in the next several paragraphs relies heavily on that work. (Freer 1986a–j, 1987a–b). (Note that the *ICAO Bulletin* changed its title to *ICAO Journal* in 1990, and that the online articles require the installation of the free DJView plug-in.)

the development of international aviation and WWII, these were matters that were left to discussion between airlines and foreign governments. This was considered unacceptable by various governments. Leading up to the Chicago Convention, national governments in the US, the United Kingdom (UK), and Canada had each independently decided that there was a need to establish some kind of international agreement. The US proposed the global meeting. Of note, of the 55 countries that were invited to the talks, 54 agreed to participate.

The three prime movers outlined their positions, and presented them to the larger group. With their own industry interests in mind, the US and the UK took fundamentally opposite views: the Americans wanted a fairly unrestricted agreement permitting wide access that would serve countries' respective air commerce interests. The British favored the establishment of a multinational regulator that would decide all things related to market access, fares, and capacity. The Canadian position was (characteristically!) a compromise, but essentially similar to the British formula. When these three disparate positions were advanced to the rest of the delegations, it became clear that there was little agreement on a way forward. Any country whose air transport industry was nascent, small, or had been badly affected by the global conflict regarded a liberal agreement as inherently unstable and something that would allow larger airlines based in economically powerful countries (the US) to dominate markets and obliterate competition. Such domination might be tolerable in other spheres of a country's industry and commerce, but not when it came to this extremely valuable and infrastructural capacity. The formulation of a working document was difficult.

Compounding the problem, air law was a relatively new discipline with little to draw upon in terms of precedent. Maritime law, which supports fairly liberal freedoms of passage, has different traditions and context, since it essentially affects only those states that can support a shipping industry, and even the most liberal and unfettered sea rules cannot change the fact that ocean transport generally only reaches coastlines. By contrast, direct air access can exist in and have commercial implications for any region or urban center of a country, and, as regards security, overflight constitutes intimate proximity to any and all of a country's centers of population, defense, and strategically or economically important commerce and industry.

International Agreements Versus National Interests

Chicago was, in a significant way, the world's opening gambit in establishing norms and codes for the conduct of international air operations, and it is unfair to call it a failure. Leading up to the conference, its goal was assumed to be a global agreement on how to determine route access, authority, and capacity—this was probably never achievable. Leaving matters of national security aside for the moment, states tend to guard air rights jealously due to the commercial (and ultimately economic) value of the activity itself. Unlike cheap surface transportation, providing expensive air transport service becomes a sizable and important part of the economies of those countries that house domestic and international airlines and perhaps air equipment manufacturers. But secure access to and authority over critical air connections is of significant economic importance to almost all countries, even those without their own airlines or aircraft manufacturers. No country wants to lose control of this valuable activity, and inclusive global accords cannot, therefore, readily serve state commercial (and strategic) interest in a way that satisfies everyone. If an agreement is liberal, and large carriers from wealthy nations do eventually dominate, their decisions about markets, capacity, and price would not be based on the particular needs of the smaller or economically weaker countries into which they may or may not decide to operate; these smaller or poorer states (with less powerful economies and airline industries) could thus effectively lose control over capacity to serve their own domestic and international markets, and suffer the economic, social, strategic, and political consequences of having substandard air service into and within their jurisdictions. Consequently, these less-wealthy or less-developed countries—and they are the majority—will never want to support an unrestrictive agreement. More powerful or better-developed countries, on the other hand, can often contemplate liberal airline agreements because their airline sectors are likely to flourish. Strong economies will therefore not support global agreements that are too rigid and restrictive or that put fares and levels of service into the hands of a global regulator, as this would compromise their ability to serve their markets efficiently, profitably (in terms of their own economic interests), and on terms that they favor in other respects as well.

As the Chicago discussions demonstrated, when restrictive agreements that control route allocation, fares, and capacity do not work for one group, and liberal agreements that open up air travel to unconstrained free market action do not work for anyone else, it is necessary and far easier to negotiate acceptable terms, one on one, with air commerce partner countries.

Now, this is precisely the matter that brackets what we can do in imposition of an international agreement on air emissions. Where commercial interests are general and neutral, all airlines must be treated equally. Where commercial interests become particular, rights must be negotiated between each pair of countries. Any attempt to align an airline's international flight emissions obligations with the CBDR status of the country that houses the airline constitutes an effective difference in commercial consequence for that airline vis-à-vis airlines from other countries that might compete with it on a route. The world's states have never been able to accept that.

Bilateral Agreements and the Five Freedoms

Long after the Chicago conference was supposed to be adjourned, the realities were reflected in the accords that parties *were* able to support: the generally accepted *International Air Services Transit Agreement* (*IASTA*) (International Civil Aviation Organization 1944b), and the somewhat more problematic *International Air Transport Agreement* (International Civil Aviation Organization 1944c), which contained the (now) inevitable provisions that vested complete control over a country's airspace and landing rights with that country's government. Other countries would never receive such authority automatically and no supranational body would make or impose decisions in that regard.

IASTA set out certain basic rights that would be routinely granted, though commercial rights would be negotiated bilaterally. The rights of overflight, landing, and commerce are called 'freedoms' in the air agreements vernacular. The right to fly over a country or to land there for technical reasons (such as refueling) are referred to as 'transit freedoms' or the First and Second Freedoms; these were set out in *IASTA*. It may be surprising to learn that even for these rather straightforward non-commercial rights to merely traverse a country's airspace or to land for fuel, there was no broad acceptance, and some countries either refused to sign or, in subsequent years, withdrew their commitment.

There was no general agreement at all on the commercially valuable Third, Fourth, or Fifth Freedoms, which were outlined (in addition to the First and Second Freedoms) in the *International Air Transport Agreement* (sometimes referred to as the *Five Freedoms Agreement*) and which involved rights for revenue service by a carrier to, from, or beyond a country other than its home. Here, everything would be negotiated bilaterally: the right to provide commercial service from one's own country to another (third freedom); the right to provide commercial service from another country to one's own (fourth freedom); and the right to provide commercial service between two foreign states when the flight originates or ends in one's own state (fifth freedom) (Diederiks-Verschoor 2006).

The authority to provide all of these commercially important services would be subject to terms negotiated on a case-by-case basis between the two countries involved, and it is interesting to examine the inevitable dynamic that resulted. Lissitzyn (1964), writing 20 years after Chicago, offers a history of how the Convention functioned in terms of the creation and evolution of bilateral air agreements (bilats), which depend on balancing respective national interests. There is no question that bilateral agreements are supposed to primarily concern the matters of needed capacity and fair share of benefit on a given route or between a country pair and, tacitly, will include some advantage for one's own air carriers. In other words, each country enters into initial discussions or review of bilats with the intention of negotiating the best deal for itself in the context of *all* of the factors that affect the perceived economic benefit of an air services agreement on both sides.

So now, let us think again about what this means from the point of view of a selective and hypothetical application of a carbon emissions sanction—one in which a carrier from a wealthy country pays a certain extra cost or bears a penalty but a carrier from a poorer country does not. Proposals for regulating emissions in international air service usually involve market-based measures (MBMs), such as a tax or trading of expensive emissions credits, for instance. If charges or restrictions are imposed upon one nation's air carriers through a mechanism external to the bilateral agreement, the balance of all of the factors that originally informed the discussions around that bilateral agreement would be affected, and the value of the imbalance would form part of the assessment of each country in *further* bilateral talks. To compensate for the disparity, the country whose air carrier had the more severe burden would demand (compensatory) value in some other way and renegotiate the provisions of market access in that context, probably to the detriment, in other respects, of the interests of the second country and its airlines. Haggling never disregards matters of economic significance, and agreements are subject to constant renegotiation. It is not realistic to assume that a party can be convinced to accept a one-sided cost factor, and negotiate as if it did not exist. That is the important point about air rights agreements that many would like to ignore today. It seems unlikely that a developed-world government would take an overt position of flaunting power with the goal of frustrating the provisions of a broadly subscribed international agreement on limiting air travel emissions; however, it *would* maneuver to achieve balance.

This frustrates those who want to apply CBDR to international flying. I point it out now in some detail, because it still pops up as an awkward barrier to an international air emissions agreement, *currently*. We will come to it again in a later chapter, but we have to understand it a little bit now. Because, in the face of Chicago, or in confronting the same issues that were raised in those negotiations and that would crop up again if there were an attempt to re-jig Chicago, there are very few options. In attaching a cost or penalty to international flying, perhaps the only real way of applying CBDR in a manner that would not upset extant bilaterals would be to exempt city pair routes where at least one location was within a state that was intended to receive CBDR protection. That would unfortunately also substantially reduce the effect of any such penalty provision.

Chicago marked a coming of age for aviation in so many ways. The broad realization by governments that their strategic and economic interests were tied closely to the development of the air industry forced a fairly comprehensive commitment to international standardization of technical elements of air operations, navigation, and traffic services; this has proved to be immensely important, as the delegations to Chicago knew that it would be. On the commercial front, the days when an individual airline could negotiate directly with a foreign government for rights to fly over and into that other country were understood to be long gone; the whole effort was mounted *because* states recognized that their interests went beyond the interests of the particular commercial enterprises that aspired to serve markets. But the lack of comprehensive, international, commercial authority deprives us of a way of assigning carbon 'blame' selectively.

International Civil Aviation Organization

One important outcome of the 1944 Chicago Convention was the establishment of a body to oversee international civil aviation matters. The International Civil Aviation Organization was rolled into the United Nations (UN) when the latter was constituted in 1947 and has been a part of it ever since. ICAO has a long record of working out standards and practices for aviation and has played the major role in coordinating aviation's efforts in important environmental matters such as noise and local air quality, which includes emissions levels. As noted below, ICAO is named in Article 2.2 of the Kyoto Protocol as one of the bodies through which Annex 1 countries will work as they find ways of limiting greenhouse gas emissions. But ICAO, it should be remembered, is not a 'from-above' regulatory body in the usual sense; it is comprised of and controlled by national delegations, which reflect the interests and prejudices that the same states make manifest in all other forums. ICAO talks have been extremely useful but have not, at this writing, established consensus on how air emissions should be addressed.

From the Five Freedoms to Emissions Control: the Kyoto Protocol

We have mentioned the United Nations Framework Convention on Climate Change (UNFCCC) and Kyoto, but, as with Chicago, let us see them in a little more detail. When the delegates to the 1944 Chicago Convention were struggling with regulations for the rapidly developing commercial aviation industry, greenhouse gases and carbon emissions were not significantly on the agenda. However, by the time the nations of the world gathered in Rio de Janeiro in 1992 for the United Nations Conference on Environment and Development (UNCED), they were well aware that our energy regime, relying on retrieving and releasing carbon that had been buried for millions of years, was a serious threat to a normal climatic evolution. And they knew that if we allowed the emissions to continue, our societies, economies, cultures, and the people who populated and constituted them would suffer disruption and destruction. And they knew that the air transport industry was deeply implicated in these energy issues. Virtually unanimous in their concerns, they produced the UNFCCC. In the Framework Convention, countries commit themselves to act to guard against the worst effects of the climate change[3] that is being caused by anthropogenic global warming. The threshold may be somewhat arbitrary, but there is nothing lax or vague in the Rio document's description of the issue and the actions required to address it, including the need to acknowledge the relative disparities between developed and developing countries in their contribution to and ability to address the problem. That is why subsequent negotiations on climate issues have taken CBDR into account; the UNFCCC spelled out the need (United Nations 1992).

Notwithstanding the fact that more recent talks have failed to get agreement on specific measures for tackling global warming, all of the information that is needed to address the problem is in the UNFCCC. The signing countries (virtually

---

3   'Worst effects' are defined by judgment call, and currently assessed as a 2°C rise in global temperature.

every country in the world, including reluctant developed ones such as the US and Canada) agreed to do exactly what was required, and none have recanted.

After two subsequent Conferences of the Parties (COPs) to the UNFCCC, there finally developed a broad (though incomplete) consensus on how to proceed with reducing and reporting emissions, including the concept, already described, of separate treatment of emissions from international marine and aviation transport (United Nations 1995). In 1998, at COP–3 in Kyoto, Japan, a document was signed by virtually all of the major emitters. The document, the *Kyoto Protocol*, set out the reductions that each country would have to meet during a specified initial period, and described how they would commit to addressing the problem.

The particular methodologies and approach for accounting of domestic emissions and what international emissions would be segregated were outlined. It is at Article 2.2 of the Kyoto Protocol that the specific exclusion of emissions resulting from the burning of fuel in (international) aviation and marine transport is spelled out, and responsibility for reducing air industry emissions is assigned:

> 2. The Parties included in Annex I shall pursue limitation or reduction of emissions of greenhouse gases not controlled by the Montreal Protocol from aviation and marine bunker fuels, working through the International Civil Aviation Organization and the International Maritime Organization, respectively (United Nations 1998).

Further, Annex 1 countries are collectively responsible for determining ways to limit international air travel emissions. Nothing about these general provisions of Kyoto is different in principle from the general provisions of the UNFCCC. Furthermore, while some states have held off accepting the specifics as covered in Kyoto, they have not rejected their commitments under the terms of the UNFCCC. So the CBDR principle and the principle that international aviation is separate have not been contested.

Common But Differentiated Responsibilities

The segregation of international transport sectors is driven, in large part, by that provision that the world's nations accepted without objection in 1992 and to which we have referred. And it remains true that developing countries cannot be held to the same standard of responsibility as the developed countries for lowering emissions. But time passes. It has subsequently become apparent that the reason for this principle is not clear to everyone who discusses these matters *now*, nor why the idea was universally accepted then. As a consequence, some do not see the reason for continuing with CBDR more than 20 years after Rio. As we continue, we see that applying CBDR is, in some ways, problematic to the goal of creating aviation emissions policy. But letting the argument proceed on the basis that CBDR was wrong or invalid in the first place will not help.

Many seem to believe that this discrimination is based on developing countries' inability to withstand the economic pain associated with reducing emissions; that might well be valid in its own right, but it is not the real reason. The point resides not in charity but in the logic of simple arithmetic. CBDR was not intended simply to address wealth disparity in the sense of aid or relief. It was accepted by the representatives of developed-world countries because it recognized a very real and logically comprehensible difference in responsibility and obligation.

In Chapter 1, we made reference to GHG residency times. Let us look at that phenomenon more closely now. Carbon dioxide (CO_2), the most important contributing substance to anthropogenic global warming, is a persistent gas in the atmosphere. Removal of CO_2 from the air is complex and dependent upon processes in the biosphere that may themselves respond to changing levels of CO_2, but an approximation of the lifetime of fossil fuel CO_2 is '300 years, plus 25% that lasts forever' (Archer 2005). The accuracy of that characterization is not as important as the validity of the general impression that it conveys. Atmospheric level is of course raised by current emissions, but current activity is merely adding to the excess mass that has accumulated since the dawn of the industrial age. And this large stock of CO_2 was built by the countries that we now regard as developed: the quantity of GHG that is now in the atmosphere was mainly produced by countries that entered the industrial revolution in the 1700s or 1800s.

Even rapidly developing countries like China, with their immense current emissions, are not responsible for the bulk of the CO_2 that is up there; so developing countries cannot be held responsible for the dilemma in which we now find ourselves. Let us focus on China for a moment because both its level of emissions and *growth* in that rate are high. China will absolutely have to get carbon emissions under control. But while it is unreasonable for China to expect that it can produce emissions until its per capita share of accumulated GHG is as high as, say, the US share (global warming and the resultant climate change would be beyond control), it is also unreasonable to insist that China and other developing countries act as immediately and as quickly as the developed nations. The converse puts the point more clearly: developed countries must accept the fact that their obligations should most immediately be addressed. The CBDR concept is a way of saying that *all* countries must face the need to bring emissions under control but that the (newly realized) urgency should hit hardest upon those countries that have derived historical benefit from and are most responsible for generating the *existing* high level of atmospheric carbon. This formulation is intended to embody the idea that the developed world would absorb the shock and take the lead in establishing ways and means by which the carbon battle could be fought by *all*.

However, claiming unfairness that countries like China are not yet committed to large per capita emissions reductions, much of the developed world has sat somewhat idle and accomplished relatively little. And the lack of action on the part of developed nations has now made it more difficult to encourage developing nations to accept restrictive emissions standards—a vicious circle that serves the interests of those who would do nothing at all.

Aviation and CBDR

Aviation's first job is to perhaps consider itself as a definitive component of the 'developed world' in its own collective mind. The Kyoto Protocol is now over 17 years old and we have passed the end of the first commitment period. We have noted the enormous problems with Kyoto resulting in failure in commitment or ratification. But, again, the one reality that was established in Rio, elaborated at Kyoto and has not been contested by anyone, is the domestic/international split in emissions accountability for aviation and marine transport. And while many countries complain that Kyoto is not fair or workable, the most seemingly irksome parts are those to which these same countries committed themselves long before Kyoto. It is true that the UNFCCC does not put hard numbers on emissions reduction obligations, but many in the developed world are not simply disagreeing with numbers, they now want to reject the principle that they accepted wholeheartedly when they initially examined, honestly and earnestly, the prospects of a climate-compromised world. So we should all remember that CBDR and the related idea that Annex 1 countries have to take the lead on developing the required low-carbon technologies (including sustainable fuel, I would now argue) were well subscribed and embedded in everyone's thinking long before the hotel rooms were booked in Kyoto. CBDR and Annex 1 obligations may be problematic in contemplation of the challenge, and CBDR operates at cross purposes to the bilats in which all countries are engaged; there are valid reasons for resisting the heedless application of CBDR provisions to international aviation. But the discussion becomes hopelessly confused when parties start to argue that CBDR was never a valid idea.

Emissions from International Transport (Bunker Fuels)

So, international aviation and maritime commerce are orphans, and they *must* be orphans, when it comes to the provisions of agreements on GHG emissions. The reasons, seen in Chicago's failure, are as valid today as ever. Everything reflects that reality. The leading scientific body for assessing climate change is the Intergovernmental Panel on Climate Change (IPCC); its 2006 *IPCC Guidelines for National Greenhouse Gas Inventories, Volume 2: Energy*, describes at section 3.6.1 the methodology that should be followed in accounting for civil aviation emissions, and includes the following paragraphs:

> For the purpose of the emissions inventory, a distinction is made between domestic and international aviation, and it is *good practice* to report under the source categories listed in Table 3.6.1.

> All emissions from fuels used for international aviation (bunkers) and multilateral operations pursuant to the Charter of UN are to be excluded from

national totals, and reported separately as memo items (Intergovernmental Panel on Climate Change 2006).

The difficulty that attends international aviation's particular status in emissions agreements constitutes a very perplexing problem. But a way around it (discussed at greater length in a later chapter) would be for developing countries to accept their Chicago obligations of being treated in a completely impartial way in the application of rules on carbon restrictions, while Annex 1 countries (essentially members of the Organisation for Economic Co-operation and Development, or OECD), would accept their responsibility for putting the technical solutions to the problem in everyone's hands. The way forward implies some collaborative commitment by developed nations to formulate policies that support the technical means of meeting those responsibilities. Ultimately, the level of emissions from any activity in any country has to approach zero (more on that later.) That will not happen if we do not have the tools.

So let us all stop for a moment and realize the vital, pressing, pivotal point in the subject we are studying here: if there are no emissions, emissions obligations are irrelevant. As we will see in later chapters, in the international arena, we are currently wrestling with the question of how to punish emissions equitably while we stumble over or ignore the questions about how to arrive at the obvious end need—eliminating the accursed emissions altogether.

European Union Emissions Trading System (EU ETS)

In the meantime, as we struggle with how the international community can agree on a system to impose sanctions on individual airlines from any country, we must face the obverse: if international aviation, treated as a separate and comprehensive entity, cannot be broken down in a way that allows one country's carriers to become subject to rules that differ from those applied to another country's carriers, can any single country or larger jurisdiction apply emissions sanctions against *all* airlines? The question has great currency since there is an active attempt to regulate in this way.

If we accept that international aviation can be treated as a separate sector in climate talks, it starts to resemble a 'country'—and a highly developed country at that. Inevitably, where actual highly developed countries have invested heavily in order to bring down their national emissions profiles, air travel is seen as an area that is not doing its part to reduce emissions. Perhaps, in such developed countries, the pace of domestic flight's attempts to address emissions might appear to lag in comparison with other costly national efforts, while emissions from international aviation might seem to be off the radar entirely. A fear or resentment of a 'free pass for flying' phenomenon may develop. (In fact, as we have seen, the airline industry has accomplished more than most in reducing capacity-specific and revenue-specific fuel consumption, nevertheless, emissions continue to rise.) So when developed countries undertake to reduce their overall national emissions amounts, they feel compelled to require both their domestic air industries (as part of their national budgets) and the international aviation sector to do the same. This is entirely appropriate. And, beyond countries, the EU, wanting to strengthen the emissions-reduction effort, has proposed in its collectivity to impose just such a requirement, though the exact formulation of the application of MBMs to aviation through the already existing EU Emissions Trading Scheme (EU ETS) is currently in flux (it had been decided at one point to make all air traffic into and out of any European airport subject to terms of the ETS).

It is an interesting example. Two important questions arise:

- The legal question: Is there anything in the Chicago-Kyoto context that might prevent them from doing this?
- The logical question: Does the EU ETS really advance the goal of reducing aviation emissions?

The legal question is being addressed currently (if not conclusively), as the imposition of provisions of the EU ETS on international aviation is under challenge and discussion on a few fronts (and temporarily in abeyance) at this writing. So we will see. But the second question, whether the EU ETS inclusion of aviation actually helps in reducing aviation emissions, is more difficult. Significantly, the provisions for inclusion in this trading scheme do not address two important points:

- the fact that Annex 1 countries have already been assigned a responsibility for reducing aviation emissions, and
- if the trading scheme excludes airlines based in non-Annex 1 countries, this would create a differentiation that would be in conflict with Chicago's axiom of uniform treatment.

There is a *de minimus* provision that discriminates on the basis of flight frequency, which roughly aligns with the size of the economy of the airline's home country, but it creates notable anomalies such as loopholes for very profitable airlines based in small, wealthy countries, and no exemption for airlines from large but poor countries whose airlines serve Europe frequently.

And this all leaves aside the fact that while the EU is a supranational entity, the bilateral air service agreements that affect its members were negotiated individually between EU countries and non-EU countries. Including aviation within the EU ETS makes it more difficult to develop broad international agreements. So, even if the inclusion of aviation in the EU ETS reduces air emissions, is it the *best* way? Should there not be a mechanism that is less complicated, problematic, and contentious? Of course there should. But many in the EU point out that we are taking too long to get to a broader, international, and workable agreement. In any event, the point that seems most relevant as regards the EU case or any of the policy difficulties upon which we have touched (and it is good to think about this early on) is that the aviation community is probably best served by figuring out a solution of the problem *itself*.

Other Reasons for Action

It would be unwise to anticipate too much the material that will be covered in later chapters on policy development. But it would not hurt to register some likely conclusions. Countries may have political reasons for wanting to achieve emissions reduction progress. But these motives are complicated by internal debate and instability as well as comparison and alignment with other countries, some of which are important trade partners. This kind of preoccupation with factors that are bound to be messy does not serve the interests of an industry that would like to resolve the matter for its own reasons. And it does not serve the interests of countries that are not looking hard enough at the details of their own interests. Not only should the airline industry itself take initiative, countries for whom aviation is currently disproportionately important should realize the advantage to be gained by supporting efforts to get their air operations onto a more solid footing through proactively developing alternative, secure, stably priced, and low-carbon sources of fuel. Eventually, migration of these new fuel technologies to airlines in less-developed countries would encourage others to get on board, as they would then have the tools to do so.

Many countries do understand how much is at stake in the effort to bring forward new sources of fuel. Countries that should understand (whether they do or not) fall into four main groups: geographically large, developed countries with a heavy reliance upon air travel such as the US, Canada, and Australia; developed countries that are geographically isolated, such as New Zealand; geographically large developing countries, such as China, India, and Brazil; and countries for whom aircraft manufacturing and export is an important part of their economy, such as the US and Canada again, France, Germany and the UK, and to some extent Italy, Spain, Brazil, Russia, and China. That understanding needs to become more apparent and turned into action; waiting for others to move first is a recipe for national and global failure.

Aviation's circumstance offers unique perspectives in many respects. On another branch of potential policy response, it has even been suggested that if important OECD nations demonstrate resolution and action through air treaties that exact higher emissions standards in exchange for more liberal access to markets, a mass of countries would grow, seeking to join the liberalized-access 'club'. Such a case can be made for building on the 2007 US EU Open Skies Air Transport Agreement, which not only provides for increased sharing of markets, but also includes statements of intention to reduce environmental impacts (Havel and Sanchez 2012). That is beyond our scope here because it requires an understanding of the balance struck by states (and the carriers that constitute important national commercial interests) in deciding between their access to other markets and other carriers' access to theirs. Perhaps priorities have evolved since Chicago, especially in recent years. But it remains true that even if commercial agreements are left untouched, the beginnings of a global international air industry commitment to lower emissions is most easily begun by moving the largest countries to act, and *again* getting the industry itself to act.

Further, to the extent that national governments do accept the reality of climate change and their obligations under the UNFCCC, they should also accept this other hard reality: if they cannot make their compromises and come to an agreement in the shared international aviation sector, they will never be able to agree to obligations at the national level. If we cannot figure out and commit to a way to make international aviation a low-emission activity in the long term, we cannot address global warming.

But we should remember that we qualified our focus on global warming and carbon emissions as being merely the most prominent piece of aviation's sustainability puzzle. If reducing carbon is part of sustainability, and if carbon-reduction technological changes must show that they themselves can be undertaken sustainably, it is time to talk about what that would mean and how it would work.

Chapter 6
Understanding Sustainability

The Nature of the Challenge

With the following chapters, we embark on a different kind of journey. We move from the task of understanding history, technicality, and background to the challenge of giving form to something that has so far resisted it. If the mission is to make aviation more sustainable, it is essential to address the most prominent deficit in that regard, the emission of carbon from the use of fossil fuels for flight energy. Understanding how and whether fuel can be produced in a way that is, *itself*, sustainable is an integral part of that task. But in order to really figure out anything about any of this we must first pay attention to the term 'sustainable' and determine what we know and what we think about it.

Despite the vast amount that we hear every day on the subject of sustainability—perhaps *because* so much is already said about it—there must be a focus on ensuring that aviation's understanding of 'sustainability' is valid and comprehensive of features that are currently part of the discourse, and also on distinguishing our understanding from any others that render the term trite, old, meaningless, or just a public relations buzzword. And we cannot study it properly or ensure that it is relevant up and down all of aviation's value and supply chains unless we take the effort to dissect what it has come to mean in various quarters, how those understandings can be most constructively resolved, and, where appropriate, integrated.

Open Source, Open Meaning

In reading and researching for this project, what I found, very early on, is that understanding of 'sustainability' is both unresolved and ambiguous. And an important implication there is that many people will be convinced that it is *neither*; that their own understanding is the only understanding and, in any case, the most logical and probably the most conventional. That will make any appearance of a long and seemingly abstruse discussion about the concept seem superfluous, or even a bit silly. But there is actually an intense, lively, and ongoing debate about sustainability and sustainable development, as exemplified by the following references:

- Luke, T.W. 2005. Neither sustainable nor development: reconsidering sustainability in development. *Sustainable Development*, 13 (4), 228–38.

- Redclift, M. 2005. Sustainable development (1987–2005): an oxymoron comes of age. *Sustainable Development*, 13 (4), 212–27.
- Springett, D. 2005. Editorial: Critical perspectives on sustainable development. *Sustainable Development*, 13 (4), 209–11.

No matter what you think about how commonplace the idea of sustainability has become, or how well established the meaning is, I ask you to entertain the observation that the idea and its meaning are understood differently, and in a huge number of ways. And the key point of making that observation is that *above all else*, those who wish to bring a human activity toward sustainability must have the clearest, most comprehensive, and well-supported idea of it to offer, and absolutely the most thorough commitment to it. Otherwise the effort will clearly fail—it will either not *be* sustainable or (just as significant in terms of practicality) it will not be *accepted* as sustainable in all important forums.

The air industry cannot afford to fail at anything. The standards are high, the capital commitments are astronomical, the competition is fierce, and the margins are infinitesimally narrow. Strategic errors by a single carrier can sound a death knell; strategic errors on the part of the whole sector would be calamitous. The collectivity of air travel industries and actors must be absolutely certain of everything that upholds what they are undertaking. Whatever the rest of the world thinks, or knows, or does not know, and whatever the views I present here, as aviation proceeds with the sustainable fuel effort, everyone must agree on the things that are absolutely essential and make sure that their understanding and consent are bulletproof. An acquaintance of mine was seriously injured diving into a swimming pool. The intended maneuver was not routine, there was some twisting and flipping involved. But the execution of the dive was not the problem. The problem was that the diving board broke. In this sustainable fuel context, an underpinning in the form of perfect credibility is aviation's 'diving board'.

A Specific Thanks

A great deal of work has been done in trying to understand what renders human activity sustainable. We owe the people who have done that work a big debt of gratitude. In terms of my effort here, thinking deeply about something that begins to effectively permeate our culture cannot be undertaken as a solo flight. Since Chapters 6, 7, and 8 are conceptual and contemplative, and address the absence of prescriptive authority as regards sustainability, finding some starting places and some deep, mature discourse and interpretation was critical. So I would like to go beyond the acknowledgements provided in the preliminary pages and mention specifically the great degree to which several years' worth of dialogue with the following individuals has helped inform what you read here: Alison Blay-Palmer, Terry Marsden, George McRobie, Kevin Morgan and Wayne Roberts. These people cannot necessarily be expected to endorse what is said in these chapters, but discussions with them, and listening to conversations between them, constitute a signal reference for this work.

The Word Itself

We have to start somewhere: the word. If I assert that we have not reached complete consensus on how we should understand the term 'sustainability', it is not an attempt to be provocative, there *is* a lack of consensus. That creates a problem: for some, the word sustainability is losing currency because of the degree of its misappropriation in discussions about environment, social justice, economics, and commerce. On the other hand, hardly any of its various interpretations are entirely inapplicable, and we must address even erroneous (in some opinions) employment of the term if we are to make progress in using it in more relevant ways. So one aspect of the treatment here is to de-emphasize any preconceptions of what the word is supposed to mean and probe ideas about what it is coming to mean.

In any case, in a post-Gutenberg world, 'definition' is perhaps becoming a lexicographical conceit. Our exposure to the history of the world's popular use of words shows that they have their own lives. But that does not mean that we can make presumptions about the acceptability of the particular way that we would want to use any word. Any option that the air industries might assume for defining sustainability in their own industry terms for their own industry purposes has been foreclosed upon: 'Sustainability' is so likely to reflect its appropriation for social, political, economic, and environmental discourse and even sloganeering that it is too late to either accept an existing meaning (there are too many) or just set out what we want it to mean (too arbitrary and not in accord with the specific meanings extant.)

The Dictionary

A good start is to accept that dictionaries exist, and that we should not use a word in a way that belies its definitions. The more popular meanings that are intended to relate to the effect of human action on this planet are valid. But let's at least *begin* with the more general dictionary definitions so that we avoid the frustration of debate that takes us into a region entirely remote from what the word was intended to mean in the first place and in the more abstract.

Also, since we can probably all agree that the important first definition of 'sustainability' would be 'that quality of a thing that allows it to be sustained', I have taken the liberty of presenting material on the verb 'sustain'. What follows is taken from the 1971 compact edition of the *Oxford English Dictionary* (two big volumes, magnifying glass necessary). At first I thought about using a more recent reference source, but such a work would indicate more about where the definition has gone rather than where it came from. With some editing-out of the various meanings that did not seem to apply at all to the way in which the word is used on the environmental, social justice, and economic fronts, those that remain are both expansive and interesting. They include:

- the support of the effort, conduct, or cause of,
- succor, support or back up,
- uphold the validity or rightfulness of,
- support as valid, sound, correct, true, or just,
- keep in being,
- cause to continue in a certain state,
- keep or maintain at the proper level or standard,
- preserve the status of,
- keep going, keep up (an action or process),
- keep up without intermission,
- carry on,
- support life in,
- provide for the life or the bodily needs of,
- furnish with the necessaries of life,
- provide for the upkeep of,
- endure without failing or giving way,
- bear up against, withstand,
- undergo, experience, or have to submit to,
- have inflicted upon one, suffer the infliction of,
- bear to do something,
- tolerate or bear that something should be done,
- hold up or bear the weight of,
- to be adequate as a ground or basis for (*Oxford English Dictionary* 1971).

In very general terms, these definitions can be seen to deal with different things. But we can ascertain that they speak of whether something can be supported and endured, on both the physical and moral fronts. We could construct a grid of four categories of meaning for 'sustainability'. And recognizing that nothing is perfect, we might as well also note the implication that where an action does precipitate insult or injury (physical or moral, to other processes or to society) there is the assessment of whether such negative effects can be recovered from.

Big 'S' Sustainability

In general, and somewhat unwittingly, the current global discourse on 'Sustainability'—referring to whether certain parts or even the totality of human action and influence are sustainable—brings all of this together and comprehends a complex quality that indicates whether a thing *can* occur or persist and then, also, whether it *should* (re)occur or persist, based on the following:

1. the availability of the resources and conditions that are necessary for it to happen or continue, and
2. the absence of threat to other things that are affected by it, and
3. if a negative effect is present, whether that effect can be endured or recovered from, together with
4. an absence of any ethical or moral offence in its perpetuation.

Current popular usage was probably originally based on the socially and economically important idea that physical resources should be husbanded in such a way that they are able to continue to exist even as we exploit them (the concept of renewable resources). But if we limit ourselves to the physical and environmental aspects of sustainability, we will place ourselves outside of credible discourse; the discussion has gone beyond that. I will argue that a more comprehensive view has probably been adopted with good reason: failing a justice obligation can have the effect of removing a physical possibility. Each element of sustainability depends upon the other elements.

'Sustainability' is now so key to formulating the world's future and to managing our societies and their effects upon the planet that almost everyone (from individuals to states) wishes to own it, often insinuating their own interests and beliefs into the discussion. Of course attributing too many different things to one word can drive it toward meaninglessness, and that is a problem. But many of the different meanings can be connected, and even when sustainability is used incorrectly, it still reflects matters of interest that must be addressed.

All Meanings at the Same Time

In the dictionary, the word sustainability is not unique in having more than one meaning. But when we use other words that have more than one interpretation, we are likely thinking of one meaning *or* the other, depending on the context. We do not usually confuse 'table' the planar surface with 'table' the organized presentation of data on a computer screen or a piece of paper. In the case of sustainability, however, popular use can conflate definitions and apply them simultaneously. But it turns out that if we think in the context of a comprehensive way of explaining how human activity can be acceptable in all respects, this comprehensive application of all meanings at the same time actually makes sense. Some will still interpret sustainability in terms of one meaning or another; this sometimes results in people talking past one another because they are each thinking only of their particular meaning and ignoring the others. For example, economists, businesspeople, and environmentalists tend to pursue sustainability on separate tracks (none of which necessarily include any social or ethical aspects). So the time has come to understand 'big S' Sustainability in the comprehensive sense; in any constructive use of the word, we must now recognize it as meaning *all* of the qualities that it routinely identifies. If we leave out any one element, whole segments of the world's population stop listening. And when we are able to see that the term sustainability is becoming used in a way that calls upon all of its various meanings simultaneously, we must push on and see it applied that way in all circumstances, bringing a little of each meaning to bear. Most people reading here will be familiar with the breaking down of sustainability of human action into environmental, social justice, and economic components. It is necessary to treat environment, social justice, and economy as all of a piece, because each element operates in a way that depends upon the other two.

The Jet Engine Comparison

Let us use an aviation example. The thrust of a jet engine (and so the speed at which it spins) can only be increased by degrees. In an automobile, one can simply push the gas pedal to the floor; the very heavy pistons and cylinders will usually accept the higher temperatures and pressures that are created, and the engine will speed up almost instantly. By contrast, a jet engine is extremely sensitive to changes in internal temperature and pressure. Increases in fuel flow need to be applied gradually, so that temperature, pressure, airflow, and turbine speed can each in turn respond to the changes and make themselves apparent so that more fuel can be metered. Failure to maintain balance will result in wild fluctuations, turbulence, stalling, and damage. Acceleration of a jet engine is managed by monitoring internal temperatures, pressures, and speeds simultaneously, so that fuel flow is increased at a pace that can be accommodated. It is not enough to manage any single one of fuel flow, temperature, pressure, and speed separately. They all respond to and support each other and they all must be managed in a way that enables appropriate interaction. Indeed, this is an obscure, highly technical example, but the environment, human society and the economy, taken together as a unit, are similar to a jet engine in that they are complex, and (importantly) operate on processes that, in each case, balance one thing against another.

The value of this analogy is to recognize that the goal is to spin the engine smoothly and increase pace at a rate that is tolerable. Managing our signals and inputs—positive and negative—together with understanding and judging the outputs is what allows the system to function. Here is another insight that the example might offer: It is not really practical for a human to watch all of the engine's parameters at the same time and adjust fuel flow to cater for each; we have to build an automatic fuel control function that *can* watch all the values. In a similar way, while it is up to humans and human observation to watch all of the parameters of sustainability, such monitoring must be put in motion with a way of automatically being sensitive to and catering for the things that are required to make that activity a long-term possibility; we cannot simply wait for problems to crop up. We know that we cannot overtly 'manage' either the physical planet or global society, but we have to have systems of assessment, governance, and policy that comprehend the changes that we make and the influences that we impose, along with a constant examination of the result. We cannot focus on just one input or outcome for long; the whole operation must either accelerate smoothly or fall apart.

The discussion really becomes one that concerns what human action 'means' in terms of effect and implication rather than what sustainability 'means' in terms of words and definitions. We have to determine how we decide whether an action is both physically and morally endurable and supportable. That is the first step toward a comprehensive understanding of whether the human project or any component part of it is perpetuable. The material in this chapter and the next two concerns the idea of sustainability of human action. Later, we will describe a major initiative in assessing and certifying sustainability in the area of alternative fuels. Taken together, these two measures—understanding and then assessing sustainability—are a lot for global society to digest. But first, we will look at historical background and context.

Sustainability's Launch

In the developed world, influences such as Rachel Carson's 1962 book *Silent Spring* (Carson 1962) and David Suzuki's writings and broadcasts from the mid-1960s onward (notably the television program *The Nature of Things*) brought environmental concerns into public consciousness. At the same time, the political idealism and unrest that characterized Western societies in the 1960s perhaps grew into an increasing awareness of disparities in human, economic and social conditions. By the 1970s, persuasive popular books such as *Limits to Growth* (Meadows 1972) and *Diet for a Small Planet* (Lappé 1975) were adding to these ideas; at any rate, the words 'sustainable' and 'sustainability' (as meaning sustainability of human action) began to enter into the common lexicon. Concurrently, and probably partially as a result, governments of the world were starting to look at ways of addressing the development deficit suffered by the poorest people, particularly in developing countries.

Considering the enormous impact—good, bad, environmental, social, and economic—of the intense development that had taken place over the previous couple of centuries, it was becoming apparent that efforts being made to reap the benefit of human and economic progress would also need to virtually eliminate its negative aspects (principally environmental, it was thought). The United Nations (UN) General Assembly recognized that much needed to be done in this regard, and in 1983 that body created a commission to examine the problem. The mandate of the World Commission on Environment and Development (WCED, known later as the 'Brundtland Commission') was laid out in a General Assembly *Resolution* (United Nations General Assembly 1983) that included the term 'sustainable development'. In the Commission's final report, submitted in 1987 and entitled *Our Common Future*, chairman Gro Harland Brundtland emphasized that the world needed 'a new era of economic growth—growth that is forceful and at the same time socially and environmentally sustainable' (World Commission on Environment and Development 1987).

That development/environment combination became a policy preoccupation and also became the most common theme for entry of the subject of sustainability into the public consciousness, as 'sustainable development'. In the process, the groundwork was laid for that concept's key elements: spreading the economic benefit of development, as a matter of human need and fairness, in a way that addressed every country's concerns, and did so without harm to the environment. So the thread of logic was (and still is) that *to be fair and equitable, economic and personal advancement* needs to be enabled, and the physical development to support such advancement needs to be *environmentally benign*. Originally, it was argued that economic development would automatically also satisfy social justice goals, but it soon became clear that this was not the case; an economy could advance in a way that was not just. Social, environmental, and economic aspects all needed to be considered individually; these became represented as the 'three-legged stool' of sustainable development.

The Three-legged Stool

It also eventually became evident that 'individually' could not mean 'separately.' But that did take some time and adjustment. It is very easy to see the abstract goals and the UN's presentation of them as being three distinct targets that could be pursued separately even if perhaps concurrently. It is not obvious that they are actually three parts of one target, and must be pursued *together*. We will come back to Brundtland and *Our Common Future* in Chapter 8 because it is relevant to other aspects of our discussion. But right now, let us examine the nuts and bolts of what the sustainability discussion has become.

At one time, our pattern of thought might have been along the lines of: local action, local effect; global action, global effect. Simplistic framing has made it difficult to see the effects of any action *in all dimensions* and *at all scales, simultaneously*. For example, thinking about social and fairness issues when contemplating a new initiative, we might at first assume that simply having buy-in from those directly involved is enough. But what about the social consequence of an environmental effect? If we destroy a local resource, that impacts justice at a local level. Any local buy-in may turn out to have been contingent, and disappears when other outcomes are understood. And what about social justice beyond the local scale? If we treat people in another area in an unfair way, as a side effect, they may feel economic pressure to exploit a resource unwisely, and the downstream result of their poor environmental husbandry is to make our lives miserable.

So we should be starting to realize that nothing is entirely local and nothing is limited in the dimensions of its effect; we need to consider the full scope of the impact of an action, not merely the view out to as far as we find convenient to consider. The habit of looking only at local effects thwarts us and takes many forms other than preoccupation with 'local' in geographical terms: local can mean 'me' or 'my company'. A corporation might talk about becoming sustainable in terms of using less of the local water, polluting the local air and water resource less, and being fair to (local) customers and workers. But from where and under what conditions does the corporation obtain its source materials? Does it import them? How? And what about by-products such as carbon dioxide, which are not usually considered 'pollutants' at the local scale? This idea of focusing on supply chains is starting to gain currency, and it is powerfully resonant in the case of aviation, where the manner of sourcing fuel (which represents about 30 or 40 percent of operating costs) is becoming the focus of the world's view of flying's viability.

We are learning more about the interconnectedness of actions. As environment, economy, and justice are carefully examined, each thread of effect needs to be followed to its very end in order to see if we are oriented toward an ideal. As we do this, we realize that when we plan a new initiative, we must address the social justice issues that may arise at scale. A project that seeks to advance the development prospects of a certain population must enjoy the physical support of not only local but also relevant global environmental and social elements.

The stance of the UN General Assembly in the early 1980s gave us a concept that has evolved, and is now swirling in the global discourse that presents itself as the discussion on sustainability. Originally we wanted development with the qualification that it must be environmentally feasible. Now we realize that there are environmental goals that are important to some people, and other goals that are important to all people, but that no goals will be realized unless development, as matter of human fairness, brings to the table all of the people who need to be involved. The environmental benignity, social justice, and economic practicality are all considered together as essential, interdependent parts of one thing. In the typical understanding of the three-legged stool metaphor, if any of the legs are absent or incomplete or faulty, the stool won't stand. But it goes beyond that: if one of the legs is missing, it is no longer a stool at all.

Over the latter half of the twentieth century, we started to understand sustainability as a particular property that could be applied to all human endeavor, which obviously included anything that contributed to development, or to being developed, or to any desired activities and ways of doing them that met the criteria that we have been discussing. 'Wealthy world' inhabitants with environmental and social sensibilities have realized that enabling sustainable development implies altering their own current activities so that each becomes environmentally benign, socially just, and economically feasible for all, and then applying those properties to their future actions as well. The activities that constitute 'being developed' need to become accessible to everyone, not only to the already-wealthy. But those activities need to be accomplished in new ways.

Sustainability by Itself

It is hard to pinpoint exactly when or how 'Sustainability' was broken out from 'sustainable development' and established as a standalone overarching quality. And, in fact, many people will offer the view that sustainability and sustainable development are the same thing. (Logically, they cannot be. If we have a blue desk, 'blueness' is only one of its characteristics; 'blue' doesn't tell us anything about what 'desk' means. Similarly, even though 'sustainability' of the global human project implies development as an economic and fairness matter that *allows* humanity to be sustainable, sustainability is still just one discrete attribute of 'sustainable development'.)

Whatever the mechanism, sustainability is now understood as a word that conjoins society, environment, and economics in assessing any human undertaking, presumably embodying the overall qualities of the original three-element model of sustainable development. When discussing whether something is sustainable, we want to know not only whether it is environmentally harmful or benign, but also whether it serves, hinders or is neutral on social justice goals. Since every human endeavor is a part of something larger and is, itself, made up of components that are smaller still, assessing sustainability looks at how the parts contribute to the larger goals on both micro and macro levels. Sustainability is thus a concern on its own and a factor that bears upon everything that we do, together with the inputs and outputs of what we do.

From Stool to Filter

Originally focused, perhaps, upon the environmental concern, sustainability has become a criterion that gives expression to the idea that our individual and institutional behavior should reflect the need to do things in a way that our global and local environments *can*, and societies and economies *could*, *should*, and *would* support so that they can each and all continue. Along the way, the image of sustainability as a stool built on the support of goals of economic development, for reasons of social justice, accomplished in a manner that protected the environment, has evolved into a conception of sustainability as a kind of filter idea that could be applied to any endeavor—whether it supported *new* development or not. This filter selects for things that are, when considered *from the perspectives of both the actors and those affected*, environmentally benign, socially just, and economically feasible. That is now a prominent feature of both policy debates and conversations about collective and individual responsibility.

In Figure 6.1 we have a couple of visualizations that present sustainability in a different way. We see sustainability as an integration of elements that support each other. Where we have two of the elements together, we have what we need for the third. Where we then have all three, sustainability becomes an intrinsic property. If we imagine the envisioned structures as 'filter' elements that operate at every stage of everything that we do, we can start to see sustainability as a property that flows through all of the things that we undertake. For one particular action or idea or project to be sustainable, the things that will be needed to complete it must have sustainability as a property.

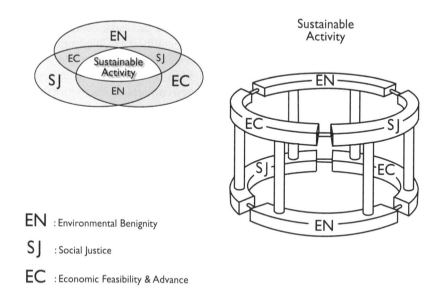

EN : Environmental Benignity

SJ : Social Justice

EC : Economic Feasibility & Advance

Figure 6.1   Two visualizations of sustainability integrating its three elements

One observation: Notwithstanding the mandate and findings of the WCED report, for example, note that there is no unanimous agreement that development is necessarily good; we will return to that.

Comes Sustainability Whence From?

Yoda (*Star Wars Episode V* 1980) taught us that being wise does not necessarily mean being clear.[1] Listen carefully, we must. Having shifted to viewing sustainability as a set of three test criteria for assessing anything, confusion ensues when people hear only the part of the message for which they are listening. The appropriation of sustainability to include different and unanticipated qualities that suit a particular want has resulted in individualized institutional ideas about

---

1   Yoda's curious speech syntax has even been the subject of analysis by academics. See Pullum (2005), for example.

the meaning of sustainability and its triple foundation. Corporate profitability, as we have noted, has sometimes been included in the thinking about economic factors, for example. That is understandable; from a manager's point of view, if a corporation is not profitable, it cannot keep going. But that fact is only relevant within the circle of parties to whom that enterprise is vital. By contrast, even though policy goals might need to permit some corporations to act in a certain way and to remain profitable in order to do so, profitability for a particular company cannot ever be said to be a necessary condition of sustainability in the largest expression of its meaning. It is just as possible that the activities that would benefit or support the larger economy, the wider human population, and the environment upon which they all depend, are incompatible with certain of a corporation's activities that are critical to its survival.

In this world, there is a tendency for the commercial aspects of an idea or entity to feature most immediately and prominently. This can distort perceptions, and we should be ready for that effect: where capital is concentrated, a powerful voice develops, a voice so loud that it causes important ideas to be drowned out. Corporate commercial viability obviously falls into a different area from considerations that are assessed at all levels, and in all dimensions including global society. We talk about sustainability in terms that are most objective, critical, and demanding. The principles operate at a higher level of concern than whether the activities of an individual or an institution or even a state are financially critical to those entities. Personal, corporate, or state matters cannot be permitted to bind the issues at a particular scale—they must *not* be bound at *any* scale. There may be a context where a certain action or undertaking is entirely sustainable, even though a particular corporation or business sector may play no useful role in it, or may even be harmed by it. For example, in some cases it may well be economically worthwhile for a state to move toward renewable sources of fuel, without any obvious way for some companies that are important to that county's economy to make money doing it. Though financially impractical for those companies, and perhaps challenging for a society, certain alternative fuel ideas are still economically sustainable; it just means that the allocation of financial resources at the larger scale is more important than the profitability or even survival of an individual business. However, those commercial voices have strength, and can forcefully influence the policy debate. It is imperative that we recognize all three elements of sustainability as essential. There is no useful, long-term, large-scale value in allowing commercial interests to push discussions through a distorted commercial sustainability lens that casts an inappropriate 'go/no go' color on everything that we consider.

Really, the goal is to find out the most sustainable way forward and pursue that regardless of advocacy on any side or any particular interest. People and organizations have to figure out where they fit.

Sustainability and Development

Though the two parts of sustainable development have been teased apart, that does not mean that they have no place together. Air travel and economic and human development are linked. The industry rightly makes the case that air travel serves and stimulates development. As a result, if the intention is to try to be sustainable, we should perhaps think more deeply about development.

One difficulty in settling our meanings for sustainability is that it can be hard to consider it in the abstract. We are in the habit of leaving the noun on the shelf and pulling down only its adjectival form when we want to qualify some specific activity, as in sustainable agriculture, forestry, fuel, transportation, development, or any number of other things. Sustainability has a venerable association with development, and this tends to monopolize the discussion. To get the gist of sustainability as a complex concept, we will re-examine in more detail its constant connection with development. But before we do, let us acknowledge another idea.

Development is Good?

It has never been universally agreed just what 'economic development' or simply 'development' signifies. And to the extent that we think we do know what it means, there is no unanimity that development is actually desirable (Luke 2005, Redclift 2005). Disregarding any judgments about this, sustainability discourse has induced a different framing of virtually everything. And it has had a bearing on how development is perceived. For example, in the view of some, it is not advisable to change a particular society's character from 'primitive' (characterized by intimate personal connection with the unmediated physical world through hunting, fishing, gathering, or relatively un-mechanized agriculture) to one more attuned to economic wealth and consumer comforts. These advocates claim that Western developed society is essentially, or even inherently, unsustainable because it is based upon practices that arose in the absence of sustainability considerations, so 'developing' is, by definition, unsustainable. That posture has its merit. At the very least, it compels us to think hard about why it might be true or not. But whether many less-developed societies were perhaps more 'sustainable' and better off in other respects is not the factor that has driven the discussion. As singer Sophie Tucker (1887–1996) famously said, 'I've been rich and I've been poor. Rich is better.' Global discourse has reflected a broad assumption of the general good of wealth by *both* many of those who had it, and many of those who desired it. And this debate about development still has current bearing—no matter what position one takes.

We do have to acknowledge that the alacrity with which the idea of sustainable development is accepted in policy and commercial spheres does not constitute any real sort of intellectual endorsement. Some perspectives posit that 'sustainable development' may be disruptively paradigm-altering, maybe oxymoronic, or perhaps even impossible: our planet has real, biophysical limits, and no amount of adjustment in the way that we do things can extend those limits. Furthermore, the social and political consequences of development, as conceived by the Western world, could halt the process even independently of the ecological constraints (Beddoe et al. 2009, Robinson 2004). Granting a place to these arguments, we proceed on the basis that we can address the potential quandary in which pursuit of sustainable development puts us. Air travel does not really have a motive or practical purpose in engaging on the topic of whether sustainable development is a useful and practical strategy. The industry's job is to understand sustainability well enough to determine how to serve policy goals in a way that *is* sustainable (whether the policies support development or anything else), and to show that its proposed remedies can be implemented and scaled in a way that allows the industry to serve markets (at whatever scale they assume) through the use of environmentally and socially benign processes, materials, and energy.

Sustainable development may prove impossible. There are voices that say that it *is* impossible. I do not say that I am absolutely certain that they are wrong. The discussion here proceeds on the assumption that development advocates and providers of developed-world services want to produce and demonstrate a way forward that allows the pursuit of goals, rather than accepting and endorsing a descent into poverty as the only alternative. Their opponents will argue that the continued pursuit of consumerist goals will make the inevitable descent quite a bit more traumatic. That discussion is beyond our scope here. The first step is to figure out what *would* be involved in meeting the challenge. With that background comment, let us get back to understanding why development has been a focus.

The Case for Development

Many people live in societies that have become pervasively and chronically poor. They live in abject and even utter misery. Those who survive into adulthood may see their children starve, suffer, and die. The levels of income (for those who have income) are so low that hundreds of millions of lives can be cast from bare subsistence into actual starvation by any blip in the price of food commodities (World Hunger Education Service 2013).

I live in a wealthy, 'developed' country as do many of the people who will read this book. It is a challenge for me to imagine what it must be like to be one of the millions who wake in the morning and spend each of their few remaining days experiencing the destruction of their societies, their cultures, their families, and their own wrecked bodies. For much of humanity, there is simply no hope without the prospect of development, so for them, and for those working on their behalf, development becomes the most important policy goal. To this population, no view of society as a sustainable entity makes sense if that view does not embody the human and economic advancement of *their* society, whose institutions *and individual members* will perhaps cease to exist without development. While some may see development itself as a key problem, holding it responsible for creating such devastating and chronic poverty in the first place, it is difficult to see how the world could now 'revelop' its way to a simpler, happier circumstance. And it is equally difficult to imagine people in either the developing or the developed world giving up on it.

But if we acknowledge that development is here to stay, economic disparities and poverty must be addressed. The entirely appropriate expectation of disadvantaged people everywhere is that no society or individual—and certainly no company or commercial sector—should afford itself the right to perpetuate unnecessary misery and death in the face of any possible alternative. *Aviation's strongest critics will reflect every particle of that expectation for as long as air industries fail to achieve leadership in the pursuit of sustainability*. So, sustainability must be part of the air industries' calculation of how flying contributes to goals like development, which is sought in a larger context for the pursuit of a viable human society.

The same people who understand the brutal realities of poverty, and even many of those who live it, seem to accept that realizing development goals for all the world's population will put truly massive pressure on our physical environment. So how do we accomplish sustainable development if it is simultaneously all of the things that form a circular relationship? I suggest that the beginning of the answer is an ability to see development as *part of* sustainability; to see 'sustainable development' not as oxymoron but as something that even embodies an element of tautology: development enables us to engage more of the world's people and institutions in the project of making the things that we do (and here is the circularity and reinforcement) environmentally benign, socially just, and economically viable and useful. That is the view that sees development as *part of* sustainability.

Sustainability Versus Development

But it is not simple. To demonstrate the complex kind of relationship that development and sustainability have, let us think about specific ways in which the purpose of sustainability can be advanced by reversing certain development initiatives. For example: Suppose there were a mine in a remote part of a country. The operation of that mine puts a harmful pollutant into the local atmosphere, fouls local water resources, exploits people unfairly, disturbs the fabric of local communities, and raises the cost of living for people in the area. The national government finds that the cost of addressing these ills is greater than all of the mine's benefits to all stakeholders—domestic shareholders, management, workers, government coffers, and everyone else who stands to gain in some way. Suppose that solving these problems were as simple as walking away: no remediation required, simply shutting down the mine resolved all of the issues. The act of closing that mine would certainly not be 'development' as we usually understand it. In fact, to use a term coined earlier in the chapter, we could call it a kind of 'revelopment': letting the elements work their will on the situation. But would this not be an initiative that nonetheless advanced sustainability? Development is only part of sustainability in a conditional way. If global humanity needs development to allow it to be sustainable, we are still left with the task of assessing whether certain development activities are, themselves, sustainable. This raises the key question for air travel: Where can we place aviation on the parallel scales of serving development needs on the one hand *and* doing so in a way that is environmentally benign and socially responsible on the other?

We can say that *everything* that makes up sustainability does so in a way that is entirely contingent. The fact that aviation supports social and economic development goals cannot be used as a rationale for sourcing fuel that is in the other specific respects not sustainable. If we need a safe, fast airplane, we cannot say, 'Well, it's not really safe at all, so we made it faster to compensate.'

Sustainable Development's Ethical Dilemmas

The development challenge and the environment challenge are both staggering, regardless of why we might take on either one. Can we do it all? Can we bring development to the poor at the same time that we are devising techniques that allow our existence to become sustainable? It is difficult enough to create solutions to the problem of the environmental load that the developed world puts upon the planet, without also figuring out how the huge balance of global population can acquire not merely 'regular' development, but this new 'wonder development'. How much flexibility do we have in the system? How far can an environmental or social or economic matter be pushed before it brings down the whole process? For example, if we feel constrained by the interdependent nature of the elements of sustainability, we may start to engage in the rather heartless calculation of whether the larger world either needs or feels morally compelled to care if certain societies and individuals exist. Currently evolving environmental difficulties will disadvantage everyone—rich and poor alike. Should the developed world just abandon the developing world and concentrate only on what it knows? This is a real problem, one that the WCED identified when it suggested that abandoning whole societies would not work in any scenario (World Commission on Environment and Development 1987).

That does not mean that everyone will accept or support a globally rational approach. Failing societies may render our collective ambitions moot, but it simply has to be acknowledged that many people just do not care. And even among some of those who do care, their actions may not give effect to any stated intention. Furthermore, since individuals, corporations, and societies all compete strategically and economically, disregard for the parameters of sustainability can become contagious. This forces us to look at the question another way: Beyond our moral or emotional reactions to certain outcomes, what is the larger practical consequence of not considering whether certain other humans or their societies live or die? Is there a values-neutral reason for doing the right thing in such situations, as the UN General Assembly's resolution implied (United Nations General Assembly 1983)? I think that there is. I imagine that most people would agree. But it takes some work to see it, and an act of will to put it ahead of some particular agenda that fills an immediate or political or commercial want. Because it acts in a longer term. It has to do with what we discussed earlier: that a society's failure constitutes not only the death of its own particular aspirations, but, eventually, everyone's. The death of societies does not occur without sucking capacity and will out of the rest. It is certain that commercial aviation, generating impact around the world by its nature, will find itself right in the middle of this issue as it evolves. If we do not act on impulses to regard all human society as worthy of attention and support, we lose the ability to *win* the attention and support of any vulnerable society, or any societies that might see themselves, in slightly altered circumstances, as under threat as well. This would create an incremental drift toward more and more failed states and communities, and it would become progressively more difficult for a globalized civilization to decide or do anything. Globalization—for better or worse—underpins everything that we do in the twenty-first century. There is simply no rational argument to support the hope for good, long-term prospects of a globalized world, unless that world incorporates the conviction that survival of each constituent society is of fundamental importance. We need our physical resources. We need economic viability. And, speaking (admittedly) very long term, in order to secure both, we need each other. If we abandon one society, we will inevitably lose another and the whole edifice will eventually crumble. Our enemy—*again*—is the shorter view. Air commerce is a small but sufficiently well-defined activity to confront the enemy, and to slay it in its bed.

Chapter 7
What Do We Owe the Future (and How Much Shall We Pay)?

If we think that we know a little bit about how human activity can be rendered properly sustainable, there is one characteristic of the endeavor that is both important and also attracts a peculiar kind of objection. Fundamental to our understanding of the word 'sustainable' is that in many cases it is used to qualify an activity *over* time as opposed to qualifying it *at a point* in time. In this chapter we examine why intertemporality is both critical to the concept as well as problematic for some.

In law, there is a general principle that the law does not work backward; some jurisdictions expressly prohibit *ex post facto* laws that retroactively alter the legal consequences of actions that took place prior to the law's enactment. Where that principle is applied, in interpreting treaty law and reparations for example, it is often used as a qualification that allows us to legally escape the results of action (including the act of legislating) that were taken in the past. But it is equally true that consequence is necessarily forward. If I chop down your tree, you are deprived of your shade only in the future. That implies an onus: we are supposed to act prudently just so that we do not have to do things like making later pathetic claims that it is now too late to fix things. Every action that we take is supposed to be forward-looking so that we do not need to try to wriggle off the hook when we look back. The law allows such wriggling in order to address matters of practicality, not real responsibility.

But we do know that such responsibility exists, and not only for actions that affect certain individuals at different times, but also whole populations that exist at different times. Otherwise, we would never even consider, for example, that reparation bestowed on descendants of members of racial or ethnic groups was justified when their ancestors were the ones harmed. But we do. Living descendants of individuals who did wrong compensate (or apologize to) the living descendants of individuals who were wronged. By extension, we will obviously be responsible if we transgress, and current and future victims may suffer. Which descendants will be owed compensation and which will feel onus all depends on how everyone acts now.

So far, we have talked about sustainability in terms of how we and other societies should act in our own fairly current, present interests. But action does have a time element. When we talk about sustainability, we are considering whether an action can be supported indefinitely and kept going; the time element respects the world's physical capacities. Necessarily, the time element also comprehends the effects of actions on future populations. So conventional understandings of the concept of sustainability do address both of these considerations. They embody the intertemporal aspect by presuming that physical resources must perpetuate, and standards of fairness must recognize and incorporate the needs and wants even of future generations. Since virtually everything changes over time, deciding whether something is sustainable becomes a complex exercise involving judgments about

1. The value of specific resources, not only in their present absolute states or quantities, but also as they may be needed or substituted for in the future, and
2. The effect of physical change on the sociocultural landscape, and how the value of any of these resources will be understood or assessed by people living in the polyglot of future cultures,
3. The effect of human action on (1), and
4. The general effect of any of this on future individuals and societies.

Comprehending people, environments, and economic benefit and viability therefore implies seeing actions and then seeing the results of those actions, understanding that results produce further results.

The pragmatist voice in our heads asks, 'How deeply can we look? How far ahead?' From an economic point of view, these are important questions, because making things sustainable can involve spending wealth in order to effect desired outcomes and circumstances, and sustainability includes economic viability and advancement as an essential element. So there is always an ongoing assessment of the degree of our financial latitude projected against the scope of the challenge. Economics is therefore wholly relevant.

More questions arise. How fast are negative effects of unsustainable behavior growing? Should we spend today's dollars on immediately starting to correct or reverse undesirable effects? Would it be better to invest those dollars in building wealth and capacity, so that more resources will be available to tackle outcomes and modification of our ways at a later date when it can be done better? Do we invest heavily in 2015 technology that will not do as good a job as 2020 technology?

Technology development, however, is an economic *process*. Like any other, it needs to start. Unless it is stimulated by market signals now, we cannot realistically hope that technology adequate to the larger task will appear at the appropriate time in the future. Unfortunately, as a matter of record, we are very slow to recognize the effects of our unsustainable actions—it has been a couple of hundred years since the industrial revolution really got underway and we are just getting around to thinking about carbon dioxide.

Benefit for Whom? When?

Another matter lurks and is a little more problematic: we are not so good at recognizing obligations to the future when we get a little selfish. Who would profit from our sustainability efforts? Discussions revolve around the long-term fate of the planet and its people; the benefits of efforts made now will accrue in the future in the form of a world of wealth and well-being greater than would otherwise happen.

Business-as-usual might make things a lot worse, so the 'benefit' is relative, but in many cases, whatever benefit that may result will fall to whatever people are living in that future. In the case of many of the sorts of improvement or amelioration that might be achieved, only a small amount of that gain falls to the average person who is alive today. Perhaps that should not matter, but it seems that it does.[1] Arguments are routinely self-serving, and economists are probably no better than the rest of us on that score.

We should discuss the whole matrix of cost, benefit, and time, arrayed against where, when, and upon whom cost and benefit fall. But we should also not allow this subject area to become a preoccupation; much of what we can do—and shifting commercial aviation to alternative, sustainable sources of fuel would be a good example—should be considered immediately valid and worthwhile in economic and financial terms. But where benefit in the shorter term is claimed to be marginal, the aviation and nascent sustainable alternative fuel industries may have to face such now/later arguments in justifying needed changes in policy or other forms of public support. It should be remembered that the industry currently has concrete goals in carbon emission reduction that stretch out over decades.

Discount Rates and the Stern Review

In terms of theoretical discourse and debate in economics, much of the criticism of immediate aggressive action on GHG emissions reduction revolves around the use of the 'discount rate'. In general, this follows what we have just mentioned. It is a discussion about cost and benefit now, measured against cost and benefit in the future: If we tie up our current capital in fighting climate change and improving sustainability, things that we could otherwise do with that capital are lost, including generating more wealth and economic capacity through other investments. This is called 'opportunity cost'. Economists address this whole topic in terms of the current versus future value of money; 'discount rate' is the indicative parameter.

The discount rate helps to decide the issue of current investment. Committing to something now for a future benefit means forgoing more immediate benefits that could be realized. So the first thing to recognize about discount rates is what is

---

1   Gardner (2006) provides a thoughtful discussion of why we find it so difficult to make the hard moral and ethical choices needed to address climate change.

being evaluated and discounted: future benefit. If we make an investment now to provide future benefit, applying a low discount rate means that we put a high value on the future benefit that we plan to reap; applying a high discount rate means that we assign a lower value to that planned future benefit than the value of what we can get right now for the same money.

In the face of climate crisis, the argument for employing a low discount rate is that there is huge future cost associated with perpetuating global warming, that an investment made in GHG mitigation is worthwhile and maintains its value— spend money now to assure the future; the future is as important as the present. The case for a high discount rate argues that tying up or spending capital now has a huge mounting opportunity cost that injures the economy immediately and does not necessarily create a sufficient benefit in the future that would warrant it. A high discount rate approach argues that we should grow the economy now; a big, healthy economy can be expected to cope with both the present and any future difficulty from such things as global warming. Focus on the known present rather than the unpredictable future. In either argument, there are important assumptions about future values and rates of return and, importantly, how much the effects of unsustainable practice will harm assets over time.

In 2005, the eminent British economist Nicholas Stern was asked to prepare a report for the UK government on the economic impact of global warming; in 2006, the *Stern Review: The Economics of Climate Change* was published, advocating strong and immediate investment in action on climate change (Stern 2006). The *Review* provoked much debate, notably amongst economists, some of whom disagreed with Stern's data, calculations or assumptions. One such critic was William Nordhaus, a respected economist from Yale University, who claimed that Stern's low discount rate accorded too much importance to the uncertain far-distant future, and suggested that the *Review* was oriented to politics rather than economics; nonetheless, he commended the report for demonstrating the links between climate change policies and economic and environmental goals (Nordhaus 2007). This does not mean that Nordhaus, Stern's most prominent (economist) critic, wants to ignore the climate crisis; far from it. He enthusiastically rebuts denials of its existence when they come to his attention (Nordhaus 2012). And he clearly supports doing something about it; in fact, on page 698 of his commentary on Stern's *Review*, Nordhaus points out that if policy measures to reduce carbon were based upon what he considers more reasonable assumptions, they would still 'go far beyond the meager policies currently in place' (Nordhaus 2007).

Others found fault with specific elements of the *Review*, but generally supported its conclusions: Harvard's Martin Weitzman, for example, also criticized Stern's rationale for insistence upon a low discount rate, but basically agreed with the *Review*'s findings on the basis of risk of very serious possible outcomes rather than on the basis of current versus future value (Weitzman 2007).

Other voices look at the very complicated matrix of factors in economic modeling and conclude that the prominent disagreement between Stern and Nordhaus stemmed very much from other factors, such as differences in assumptions about climate sensitivity and assumed rate of growth in technological capacity, in addition to discount rate (Espagne et al. 2012).

In the view of many people, these distinctions are a little too subtle; the debate seems esoteric to the point of being bizarre. I think that it is wise to take a little of this last perspective with us as we continue to think about these things. Lord Stern is a noted economist of great repute. His critics also have their credentials. But arguments seem abstruse in the extreme, given potential outcomes.

For the aviation industry, it is also important to know what common public perception and sentiment would support. Stern's arguments—even if based upon what appear to be merely theoretical considerations about discount rates—seem, to many reasonable people, to be on the right track; the findings resonate with us. One would use perhaps more conventional discounting for making personal purchase decisions. But, by implication, other decisions should comprehend discounting in ways that respond to other factors including, of course, the future of human society, if, indeed, that future is at risk.

Charts, Graphs, Economic Models, Climate Models, and Crystal Balls

No one will claim that these things are easy to assess, but it is problematic to assume that the onset of danger is relatively gradual and comprehensible. In fact it is the very lack of predictability of perturbed climate systems that would argue for the lowest discount rate. In the cases that some other economists make against Stern's position, there is rarely any reference to the reality that is evolving, or could evolve. They do not much consider any possible and absolutely unquantifiable non-linearities, or say how they could be accounted for. If a steadily rising global temperature provokes another occurrence that, itself, initiates an even more dramatic effect on temperature, our discount formulae cannot seem to capture that in a way that makes sense. The discussion implies that some economists have figured out and factored in all the things that we cannot possibly know about climate! The irony here is that unlike some species of cynics and skeptics, they seem not only to accept the findings of climate scientists, but in some cases appear to accept them too literally; where valuable but necessarily somewhat primitive climate models are misunderstood to imply steady, or at least predictable growth in global average temperature, it is perhaps easier (and more reassuring) to think that humanity will figure things out as we go along, but we do not know very much about how global warming and climate change effects will be compounded.

Perhaps other influences are in play as well. It seems that much of the argument against the low discount rate supported by economists like Stern and Weitzman reflects an ideological predisposition to see long-term low discounting as a bias against a market economy. That may be a controversial position, but it is one that some environmentalists (and presumably some airline passengers) might take. Is ideology a factor? It is not hard to make such an argument. In general, discount rates are accepted as needed to calculate a balance of near-term cost and later benefit for *current* market actors, most of whom would certainly apply discount rate theory in trying to decide what a current personal investment might return to them in the future. But calculating a later return of benefit for people who are not even born yet and *cannot* be market actors in the classic sense perhaps rubs some economists the wrong way. That motive cannot be imputed to all the economists who argue that Stern and others are misguided, but if it applies to some, that could contribute to some of the fuzziness that we experience when we try to see the economic issues clearly.

Moral Issues

Perhaps we are better served to think in terms of risk, assessing the things that we know are important to us as both individuals and, in the case of aviation, the industry's real future. Discussing the prospects for human society in terms of economic technicalities can seem arch. Ask anyone at all if they value their children's future and their grandchildren's future, and they immediately answer, 'Yes, of course.' For most of us it is easy to construct a perpetual chain of concern that implies an unambiguous commitment to the future. We care about the future of our children, and since a key concern of our children is *their* children, we therefore care a great deal about our grandchildren as well. Most people, whether or not they have descendants of their own, are willing to make sacrifices when they think of younger people and future generations suffering as a consequence of our neglect. This is the moral reality that underpins the intergenerational considerations made clear in the World Commission on Environment and Development's thinking in *Our Common Future* (1987) and in the general public discourse on sustainability. That is the way that most of the air industry's customers would look at it. But what of the industry itself? Well, the industry's most important current goal is to halve absolute emissions by 2050. That acknowledges immediate and future obligation. It is an embedded part of industry thinking.

What Are We Giving Up?

The arguments about discounting theory are esoteric and presume that we would be able to explain and justify our decisions to future generations, whether we chose to invest now in damage-mitigating initiatives or to grow wealth to hand over to our descendants. But we have not even been able to accurately assess the value of the capital that we might put at risk in any approach. And what do we say to those future generations if we destroy capital against the speculative value of some presumed rate of return for them, and our calculations are wrong? Since we are talking about the future of the entire world, arguments based on discount rate theory would make a lot more sense if they all assumed the worst possible case. Because if they *did*, the numbers might be even more cautious than Stern's.

We are constantly thinking about costs and benefits as they will be reflected in assessments of current and future capital. This capital has always been viewed as a combination of the wealth of natural resources together with the value added to the economy by human work and ingenuity. A 1997 paper in *Nature* estimated the then-current minimum value of the entire biosphere at between 16 and 54 trillion dollars, or 33 trillion dollars per annum (Costanza et al. 1997). Much of this occurs outside the market, but we never view this capital as diminishing; we assume that any resource that we extract and any innovation or work that we contribute should act to *build* that wealth. But how do we do the math when we act to *destroy* natural capital? If we extract petroleum, turn it into fuel and burn it, and in so doing also simultaneously diminish the value of the atmosphere in terms of its ability to maintain the desired temperature, how do we apply relative value to that? Whatever the calculation, natural capital stock must be given appropriate weight in decision-making processes. We do not know how much the atmosphere will be damaged and we do not know to what degree its temperature-moderating capacity will be altered. We only know that anything that we can do to adapt or compensate is uncertain in terms of cost and value.

Valuations of the 'social cost of carbon' (SCC) vary hugely according to which particular model is employed and which particular assumptions are inserted. Our ecosystem, with all of its interactions, is essentially incomprehensible.[2] We do not entirely understand how it works and know very little about its failure modes. Consequently, we do not appreciate what, in terms of quantity of benefit versus harm, we are risking or to what degree we are endangering the system. Our lack of knowledge prevents us from being able to say how future technology might be able to address any calamities that might result or threaten, so it seems silly to hold a blind assumption that 'somehow' technology will take care of all of the problems. In effect, we are risking what we currently do not know or cannot quantify against future prospects for gaining wealth or opportunities to rebuild it. And we cannot realistically presume to know very much about how those prospects or opportunities will present themselves, either.

Externalities

While we are thinking in terms of economics, we should touch on the related matters of 'market failure' and 'externalities'. To a market theorist, the power of the market is to allow market actors to perceive value and interact in the market to clear that value. To the extent that there is a real value that is generally not perceived in the marketplace, the market 'fails'. But value that is not perceived may still act to the benefit or detriment (if the value is negative) of the world at

---

2 One way to get an idea of how difficult it is to understand complex interactions would be to review the challenges encountered within artificially closed system situations such as Bios3 or Biosphere 2. See Nelson et al. (2010) and Poynter (2009) for examples.

large and all market actors. Because such boon or harm remains unaccounted, it is 'external' to market function.

If market failure and externalities are not entirely apparent to the layperson, they are not completely obscure either (Bowen et al. 2012). But the fact is that we seem to struggle in facing known market failures and attributing a value to negative externalities. Given what we have said about how economists wrestle with discount rates, it is not hard to understand why: identifying and ascribing cost to negative externalities—even though they may be understood by some—'feels' like arbitrary impost to many. That makes a charge politically challenging. And the economists will point out that, exactly like the discussion on discounting, it is difficult to agree on how much.

All of that has relevance here, because it is relatively tempting for some voices in the air industries to conjure excuses for refusing to accept the obligation to impose an appropriate cost associated with ridding the skies of carbon. When we discuss policy in following chapters, we should remember all of this, and we should particularly remember that perfect knowledge is not always necessary; doing *something* and then ramping that up as required is always a useful way of getting started.

Against all this discussion about what is happening and what we can do, there are those whose stances will not budge. They simply maintain that we cannot be absolutely certain that anthropogenic global warming is real, or how much damage we will do if it *is* real, or that we will not eventually develop technologies that will reverse the warming and its effects. Selfishness or ideology renders these detractors obdurate, resistant to scientific evidence, and willfully blind to consequence. But let us frame this as a more immediately personal metaphor: Recently examined patients cannot be certain that their doctors have identified their disease properly, nor how much illness, or disability, or risk of death they may expect. But if virtually all of the doctors diagnose the same disease and predict that it will likely become chronic and perhaps fatal, one wonders what sort of person, facing the expenditure of a relatively affordable amount for the recommended treatment, would say, 'Well, until the doctors are all absolutely certain how this is going to work out, I'm not interested in paying for the pills.'

The point is that the air industry can expect a certain portion of their customer base to lose patience with any balking or denial from any industry voice that argues that we do not need to act—or not right away. And I wanted to particularly remark on the fact that arguments that appeal to the selfishness of generations or to the denial of externalities must be understood for what they are, before we continue with the subject, because as we will see, principles of intergenerational justice are heavily relied upon.

Chapter 8
Closing the Circle

Impacts and Interactions

Our current, special use of the term 'sustainability' is a human postulation and an entirely appropriate one. By conceptualizing ideas about the long-term feasibility of those things that make up the human project, we are creating the intellectual tool that is required to *understand* the world and our place in it, and to formulate and structure our larger actions to embody an appropriate and very limited degree of disruption. Here, if we posit sustainability as an important driver of our philosophical orientation and actions that may *affect* our world and our place in it, we can gain a useful way of pursuing our continued viability in a world that works. We begin to articulate a view of a dynamically stable system. Humans are co-constituents of the physical world, part of the Earth's thin veneer of living matter, along with water, gases, minerals, and so on. Since our effect cannot be nil, we must balance our actions and outcomes constantly, always trying to achieve a presence that alters systems as minimally and as locally as possible, maintaining the dynamic stability. We bear in mind any effects at boundaries and downstream locations.

Human Actions

We have talked about economic postures that deprecate the push for action, built upon claims of either poor accounting against time and discount, or failure to make the case that the effects of carbon in the atmosphere (for example) are a legitimate market externality that should be recognized. The world's prospects should be contemplated on the basis of facts and caution. The test of human intelligence as a survival characteristic is whether we can manage our way out of the problems that we create as quickly as we are creating them. A beaver, for example, disrupts local conditions by building a pond in a local ecosystem that is adapted to such disruption. So its pond does pose some temporary challenge to individuals of other species (drowned trees may die) but assists yet others (moose obtain a perfect watery environment). If the pond becomes too big and results in its food being too far away, the beaver's nature is to address and adjust this, and so maintain a very simple balance. But whereas the beaver needs no abstract conceptualization of sustainability, *we do*. That is a key difference. Our capacity to affect things extends far beyond the local and the limited. We are capable of changing the path of everything. Taking the beaver's approach of pursuing our wants and waiting for circumstances to let us know that we have gone too far is not an option for us. If our capacity for disruption is not to prove globally destructive, we must adjust our thinking and become aggressive in analyzing the possible consequences of what we are doing, and also in remediating those consequences and the subsequent actions that they precipitated.

Short Sight

When we look for disruption of our world, unsurprisingly we naturally look first at things that most strongly, or immediately, or directly affect ourselves. That is a good instinct, probably. But it can get us stuck: the same impulse drives us to look at effects in the smallest context—*not* a good thing, since sustainability of action has no real meaning if we only apply it selfishly, locally, or selectively. One of the first tasks in bringing sustainability into practical application is to defeat this impulse.

But for many, human society is not (yet) viewed as global society. Ultimately, this attitude has real physical consequences. Sustainability actually implies assessment of human action and its effects, conditioned by the human framework of social interaction, and not limited in comprehension of interest related to scale. The discourse should comprise *all* the questions that an individual might want to ask: Can I live with this? What about my community? My organization? My country? What about the rest of the world?

All of this forms part of public discourse and affects that discourse, so we cannot talk about the idea of sustainability without constantly delving into those conversations, developing our thoughts more completely, and coming back with something that makes sense in an appropriately larger context. We have to start recognizing not only that our actions can affect us directly by affecting our environment, but also that even when our actions only affect others, their reactions and responses may in turn affect us. We have to cultivate the habit of examining our plans for all possible outcomes—physical, social, economic, political—at all scales over time.

Hardly anyone thinks about sustainability in the depth that we are considering it here. We need to know that as the decades pass, and industries like air travel review the plans they made for those decades, realization will deepen about what has to be considered when deciding whether an activity can be allowed to continue.

Too Soon Old; Too Late Smart

The matter of cascading and interacting actions and effects is certainly important. And though we must recognize a need to look far and deep in contemplation of our policies and undertakings, that does not mean guarantees. When we think about our need to understand our actions and their outcomes, we must also realize that we can and do create consequences that are impossible to foresee no matter how hard we try. We may be able to understand direct effects, but we routinely ignore the possibility of cascading effects and resulting non-linearities. Part of being wise is accepting that there are a lot of things that we do not know and will necessarily *not* be successful in anticipating.

Sustainability and Development: Global Warming is Everything?

Climate debates, summits, agreements and policies over the past couple of decades reflect the view that the developed world bears the responsibility to act first and most in creating (among other things) new low-carbon technologies and sources of energy, and is also obligated to figure out ways of putting these de facto development tools into the hands of those most urgently wanting development. This accepted precept underpinned the outcome of the United Nations Framework Convention on Climate Change (UNFCCC) negotiation of over twenty years ago. The principles of that convention were not then, and are not now, controversial; leaders of countries rich and poor, conservatives as well as progressives, all agreed on them. However, important constituents of a developed country's political establishment may still deny the reality, and political leadership cannot always bring itself to be honest in presenting the sometimes difficult changes that are really required. Unless climate policy reflects the broader and more legitimate perspectives that determine whether human action can be sustained, many societies will withdraw from the discussion and thus greatly exacerbate the difficulty of tackling the problem. Our focus with this book drifts inevitably toward climate. But there is a need to see even that enormous plight as addressable only if we remember that there are other environmental, justice, and economic issues that must become or remain properly addressed as well.

The Problems of Mix and Match

Sometimes we forget. The predominance of global warming in sustainability discussions has created a complication. We (mostly) accept that environmental, social, and economic aspects are all essential components of sustainability, but we also increasingly regard that the elements are, in some respects, severable and can be assessed one against the other within the overall context. We are also realizing that the environmental impact of global warming itself will often loom larger than each of the individual economic and social effects that we *might* create in fighting it. Unchecked, global warming will bring about unprecedented economic and social catastrophes, on every scale—including local and regional. The trap of assuming that at every scale, every element of sustainability is perfectly severable and substitutable tends to provide refuge for those who seek an excuse for failing to address other particular sustainability considerations. At the very least, it can erode focus.

From the point of view of opinion makers in the developed world, all of the foregoing, along with emotion and prejudice (or even just plain old shallow thinking) create some interesting, contrary, and problematic views. Global warming can become considered in rather contradictory terms.

First, the underlying conflict:

- Global warming is real (according to most political and intellectual leadership).
- Global warming is *not* real (according to some political and special interest and advocacy groups).

Then the conditional or qualified response:

- *If* global warming is real, tackling it is of a higher priority than addressing social and economic inequity, because it will, if unabated, create *more* inequity (true, but a simplistic characterization that does not acknowledge the development gap that was partially caused by the developed world's exploitation of fossil fuels).
- If the developed-world economies stumble, they will lose the wherewithal to stop global warming (true).

This sample of perspectives does not, of course, constitute a collective or consolidated official view of any government—let alone entire societies—but they are positions that inform whole national debates and (to the extent that sustainability aspects of global warming are acknowledged at all) engender an understanding of sustainability that concludes that the vibrancy and pace of growth of developed-world economies is more important than anything else. From the air industries' points of view, and in consideration of policy initiatives, such attitudes tend to cause fractures in the local, national, and global consensuses that will be essential to an industry that operates everywhere.

Assigning Responsibility: Per Country, Per Capita, Per Consumer

One more point of view manifests in debate and forms an important part of the policy discussion in the developed world. Combining the ambivalence described above with advanced nations' frustration when confronted by an unanticipated and unwanted problem of their own making, we hear, 'The developing world has to reduce its emissions too.' This is a path of discourse that the commercial aviation community should do everything in its power to keep the rest of the world from taking. It was a divisive and recurring theme from some of the developed-world representatives at the fifteenth UNFCCC Conference of the Parties (COP 15) meetings in Copenhagen in 2009 (Diringer et al. 2009).

Indeed, the developing world *can* limit its emissions, if its development path includes new energy technologies instead of the fossil fuel dependency that now troubles industrialized countries. But the 'them too' assertion implies that an absolutely equal national pace of change is appropriate regardless of other factors. That is quite wrong.

First, it is often clearly unfeasible; whole countries and broad regions of other countries may produce meager levels of emissions that are not reducible, and these areas simply cannot limit emissions in the same way that the developed world could. Also, by its own admission, the developed world is accountable for more of the current levels of GHG extant, and so bears greater responsibility for getting sustainability-enabling technologies and other resources, including financial, into the hands of developing-world populations *so that* they can develop without exacerbating the climate disruption.

It's Not Who We Are, It's What We DO!

In any event, it is not really countries *per se* or their governments that produce (much) carbon dioxide. It is human activities—wherever they occur—that put GHG in the air. In this respect, we (and especially those of us in the aviation business) might find it more useful to focus on actual emissions-burdensome activities rather than always thinking solely in terms of national obligations. The activities themselves are the problem, and finding ways of reducing the carbon burden of, say, manufacturing, is not very helpful if those new systems cannot be made available to everyone. On top of their inherent carbon burden, developing countries, understandably searching for the cheapest way of making things, may unintentionally create a 'leakage' problem when companies in wealthy countries send their manufacturing tasks to those who are not thinking about emissions as a priority.

Focusing on the activity rather than the jurisdiction gives us a different view. Even within developed countries we find people who contribute very little to the emissions problem. And looking at developing societies (even some that are well embarked on a development path) reveals not only whole swaths, or substantial parts, of economies that do not enjoy expanded energy use or high standards of advancement, but also regions, or centers, or individuals within those countries who *do* exploit energy and generate emissions in an extravagant way.

Thinking about activity rather than borders, consider China as a particular example. While it is true that China has demonstrated a high degree of energy innovation, it has still taken the lead among countries in terms of total *current* national rates of GHG emission and is routinely singled out as a large emitter nation that must take responsibility for its global warming profile. That is fair comment; it must. However, it is also true that China's population lives in a largely under-developed economy. The online *World Factbook*, a publication of the US Central Intelligence Agency (CIA), paints an interesting picture. Based on 2013 estimates (all figures in $US), China's national Gross Domestic Product on Purchasing Power Parity basis (GDP-PPP) is $13.39 trillion; that of the US is somewhat greater at $16.72 trillion. But per capita GDP-PPP in China is only

$9,800, which contrasts with US per capita GDP-PPP of $52,800. A national economy that was only one-fourth larger than China's supported a per capita GDP that was more than five times greater. This reflects the difference in the respective populations (2014 estimates): China 1.36 billion, US 319 million—a factor in excess of four.

In China, a relatively smaller percentage of the population is engaged in the economic activity that generates (and presumably consumes) the wealth. Fully 33.6 percent of China's national workforce is employed in agriculture, which generates a mere 10 percent of GDP. Most (90 percent) of the wealth is generated in industry and services, though only 66.4 percent of the people work in those sectors. In the US, there is a much stronger correlation between where the wealth is generated and where the population is engaged: agriculture employs 0.7 percent of the population and results in 1.1 percent of the GDP; industry and services involve 99.4 percent of the population and provide 98.9 percent of the GDP (Central Intelligence Agency (US) [continuously updated]).

This does not mean that the US does not have its own income disparity problems; within the various sectors there can be large economic discrepancies that would imply differences in personal shares of emissions. But in China, large numbers of people and entire geographic regions have little to do with the country's overall wealth generation, energy consumption, and emissions. In 2005, per capita rates of GHG emissions in China were estimated to be at about one-fourth of those in the US (International Consortium of Investigative Journalists 2009). Are there people in China who are pursuing lifestyles that involve the flagrant kind of GHG production that typifies those of wealthy people living in the developed world? Yes, but not 1.36 billion of them.

The responsibilities of the developing-world societies are not in doubt, but they are simply not on the scale *nationally* of the obligations of those parts of the world or specific activity sectors that have derived a huge economic benefit from the exploitation of fossil fuel energy over the last two centuries. It is growth in the activities that constitute being developed that raise national emissions totals in developing countries. From that perspective, it is logical and appropriate to tackle the energy and emissions of problem activities simultaneously, and even together. Any perceived recalcitrance on the part of developing nations cannot warrant any delay in moving forward wherever possible.

Rather than pursuing more futile debate about whether countries such as China should do more or less, perhaps it would be more fruitful to identify and appropriately address individual problem sectors, regardless of where those sectors might operate. This is perhaps the only way to bring limitations on intense emissions-generating activities into parts of the world that generally produce lower emissions on a simple per capita basis. And, clearly, aviation would be a case in point; while aviation's current total rate of emissions (and certainly its cumulative emissions share) is low, it *is* an industry that features the kind of energy and emissions intensity that needs to be addressed. And since international aviation has already been singled out as especially problematic, there should be wide support for attending to aviation's sustainability deficit.

Opportunities: The Possible Interplay of Development, Aviation, and Making Fuel

While we are examining distinctions between developed and developing societies, and the fact that emissions-intensive activities can be exploited anywhere, perhaps we should tie the two ideas together and just be conscious of the possibility that making things, such as fuel, that wealthy people enjoy does not necessarily need to happen solely in those places where they will be consumed in the largest quantities. Furthermore, making fuel can raise levels of economic activity and become part of building the wealth that will exploit it. The debates about how aviation can be rendered sustainable (or more sustainable) are front-loaded with political issues because flight energy has become a showcase example of hogging, by the developed world and wealthy individuals, a finite resource that also produces persistent global warming pollutants. The most central feature of the politicization of sustainability in the context of global warming is this: In the view of many, fossil fuel use has constituted an unaccounted subsidy for development and wealth. Attempts to bring the total amount of fossil fuel use to a dramatically lower level are non-starters and patently unjust if they seek to preserve the advantages that the developed world has secured (in an inappropriate way), or to force an even more difficult path upon people who are only tentatively started (or not started at all) on their way to a better and wealthier existence.

Accepting that the rate at which energy is expended in a society is indicative of at least part of that society's level of development, aviation liquid fuel constitutes an important factor in the development equation. If aviation fuel is produced in an environmentally and socially benign way, we can say that it contributes to development, generally. But if it is produced in an area that particularly needs economic advance, it could contribute directly. This is a good time to note briefly that the aviation and fuel industries should not ignore opportunities to promote sustainability and development, sometimes together. Many biomass and processing technology pathways or more direct energy-to-fuel technologies may arise, some of which might be most appropriately located in the developing world; harvesting the biomass required for some biofuel options and producing sustainable alternative fuels may well assist with development goals in poor regions. There will be *a lot* of options to explore, and OECD countries that host sustainable fuel enterprises—regardless of whether they signed the Kyoto Protocol—have to remember the very real obligation to look for ways to assist the developing world both broadly and specifically with establishing sustainable jet fuel capacity.

Promoting Sustainability: Obtaining Buy-in

But regardless of specific matters such as who makes fuel and where, the larger point to which any of this pertains is that nothing really survives questions about its long-term sustainability unless it receives the broadest support. Understanding the social and economic aspects of sustainability is crucial. If our initiatives create a situation that discourages people from acting on sustainability issues because they feel that they are doomed in any case, then those initiatives are not sustainable.

What's In It For Me?

The discussion thus far, throughout the book, implies the following: The measure of sustainability of the human project is the measure of all things, tangible and intangible, for all people, together with the ecologies that support them, over all of time. This comprehensive sense of the term is now broadly implied, and sustainability is now frequently (I might venture 'usually') discussed this way. As an ideal, it works well. But something that contains so many elements is difficult to hold together as individuals or states or corporations grab the part that interests them and run with it. Again, we have talked about state versus state, state versus activity, and activity related to individuals who exploit it. Beyond that is our earlier point about what particular parts of sustainability might be accepted by these players, and which not. Obviously, a specific, narrow, subjective conception of sustainability cannot really work when we examine it against all the features and details of energy, emissions, water, land use change (direct and indirect), social justice, economics and the whole constellation of things that really pertain. But many of these considerations are both hard to quantify and interdependent. So while it seems irrational for any entity to prioritize any single criterion, it happens; we each probably understand that we cannot think about sustainability in a personal or local way, but the desire to put one's own interests first dies hard. We know that when all factors interact and a human's effect on the world cannot be nil, the sustainability of particular items is a matter of degree and compromise; something can be either more or less sustainable in one particular dimension, at a larger scale, or to a greater degree. That tempts people to emphasize selectively.

So it *is* a challenge to engage individuals and ask them to regard sustainability as that three-legged stool, where all legs are simultaneously and equally important. When we identify big problems (such as carbon in the global atmosphere), our collective impulse *should* be to prioritize issues based upon criticality at the largest scales downward. But we have to work actively to retain support for that approach. So we also have to understand and recognize issues that are, though local, also quite critical. One of our first responsibilities is to give some thought to how sustainability elements are related to one another and to demonstrate commitment to the ones that will keep all the parties at the table. 'Selling' a sustainability perspective is a matter of identifying the matters that are of global concern to everyone and acknowledging the needs that are particular and immediately compelling to each.

Truth in (Sustainability) Advertising

Part of getting and keeping support—as in anything—is honesty. The attention that individuals and organizations are prone to pay to their particular needs is both a commonplace and a barrier. One effect of that is the prevalence of 'greenwashing'—an organization or business selling an appearance of sustainability both to stakeholders and to themselves, while holding short of or failing to see the need for real, large commitment. Many individuals and groups are at the point of trying to do at least some things in a more sustainable way, but their conceptualization efforts are varied and sometimes contradictory. They may seek ways of reducing energy consumption or waste production but do not seriously aim to be a more comprehensively sustainable operation. Bill Wallace, a respected engineer and sustainability professional, calls this 'accessorizing for sustainability' (Wallace 2012). They try to turn the pursuit of some particular improvement into an outright claim, to don a sustainability superhero cape when, instead of leaping over tall buildings, they are really just inching their way up the brick wall to the first floor window. The value of at least getting started cannot be overstated, but it is almost pointless unless we are intending to go further and make that intention clear.

People can generally recognize overblown declarations and public-relations fluff, and will examine air travel's claims about sustainability and shifts toward a 'sustainable' fuel on this basis. Accepting that there will be such a level of cynicism is perhaps easier if we remember that real people are, every day, working within, selling to, and buying from corporations—large and small—that are making claims of sustainability in any number of ways and degrees, and for any number of reasons. Halfhearted efforts and exaggerated declarations do not fool the average citizen, because as workers and consumers, they live the reality. They are aware that less rigor is being brought to bear in the application of the term 'sustainability' than is proper, and also recognize that different voices use the word to mean different things. For any company or sector to be credible in the face of jaded popular view, it must demonstrate an awareness of the public's ideas about sustainability and also deliver real, solid, relevant, and visible sustainability performance. On the basis of my experiences with both the airline industry and environmental advocacy, I offer the following observation: While many undertakings may dress up as sustainability and take themselves off to the ball, an initiative that does not advance the most comprehensive and demanding understanding of uppercase 'S' Sustainability at a global scale is going to be criticized pretty mercilessly.

The Starting Point and the Road Forward

The purpose of these few chapters has been to offer a way of regarding sustainability that will seem comprehensible, reasonable, and acceptable to most people. It is certain, however, that it will not coincide with the various ideas that already have a comfortable home in the minds of some. If there is reluctance to adopt the general view presented here, an important exercise is to trace some of the threads of ideas about sustainability in more detail and see where they came from and how they arose. We are talking more about the operation of sustainable human enterprise and how efforts in that direction are perceived, so it is probably worthwhile to back up and see where some misconceptions arose. I will again make reference to the United Nations (UN) and the World Commission on Environment and Development (WCED), also called the Brundtland Commission. A lot has happened since 1987. Let us take a critical look at this landmark document in light of what has been put forward here.

The Impact of the Brundtland Report

Referring back to the UN General Assembly's resolution establishing the World Commission on Environment and Development (United Nations General Assembly 1983), recall that the concern that dominated that body's mandate was development, and whether it could be accomplished in a way that was environmentally feasible. That mandate was restated in the Chairman's foreword to the final report, *Our Common Future* as: 'A global agenda for change' (World Commission on Environment and Development 1987). The WCED mandate was an urgent call by the General Assembly of the United Nations:

- to propose long-term environmental strategies for achieving sustainable development by the year 2000 and beyond;
- to recommend ways concern for the environment may be translated into greater co-operation among developing countries and between countries at different stages of economical and social development and lead to the achievement of common and mutually supportive objectives that take account of the interrelationships between people, resources, environment, and development;
- to consider ways and means by which the international community can deal more effectively with environment concerns; and
- to help define shared perceptions of long-term environmental issues and the appropriate efforts needed to deal successfully with the problems of protecting and enhancing the environment, a long term agenda for action during the coming decades, and aspirational goals for the world community. (World Commission on Environment and Development 1987)

The report itself is enormous and detailed, and was received with much fanfare. The following much-quoted passage appears in the introduction to the report's *Chapter 2: Toward Sustainable Development*:

> 1. Sustainable development is development that meets the needs of the present without compromising the ability of future generations to meet their own needs. It contains within it two key concepts:

> - the concept of 'needs', in particular the essential needs of the world's poor, to which overriding priority should be given; and
> - the idea of limitations imposed by the state of technology and social organization on the environment's ability to meet present and future needs. (World Commission on Environment and Development 1987)

Understandably, due to the ordinary nature of the word, and its then newly commonplace use in the context of development goals, the term 'sustainable' (and 'sustainability') was never formally defined in the report. The above passage has been appropriated as a definition of sustainable development. But, increasingly, it is taken to be the 'official' definition of sustainability itself.

It is important to understand why this is a problem. And no real blame attaches to the authors. First, this definitional piece was not really posited as 'the' definition. In context, it forms an excellent and appropriate part of running text. But since it was not really intended as the defining statement, it does not serve well in that role. However, it is used.

When it is put forward as the authoritative definition of sustainable development, one might say that it fails because it stresses only the intergenerational component of how development can be sustainable. This emphasis is appropriate in context, but in isolation and misappropriated as a complete definition, the words seem to imply an imprecise, interest-driven, and subjective approach. It can read as if it were merely another example of how understandings of sustainable development get to *be* interest-driven and subjective. The text is valid; the disappointment is with the way that other text has been ignored. And the 'Brundtland Definition' has been somewhat misused by those whose perspective led them to appropriate it as a definition—not only of sustainable development, but of sustainability itself.

The framing of the Commission's task reflected an important emphasis: development. The mandate required the Commission to describe how development (clearly important and necessary) could be achieved in an environmentally acceptable way. The General Assembly, the Commission, and others talked about 'development' that was 'sustainable'. But the conjoining of those elements as 'sustainable development' was new. It needed to be described. It was; but there is a *lot* more text in the report than that one passage. When taken as 'the' authoritative and sole definition it can be taken to imply an inapt assumption that we make about the world being required to serve us without us thinking about how that can happen. Human 'needs' in perpetuity are singled out as a key concept. That is fair enough even if the explanation of 'needs' is not present just there in that passage. Then it is claimed that the needs of the poorest should have the highest priority— also fair. But the way in which 'limitations' is used can be taken to imply that only the level and kind of technology and social organization limit our ability to extract what we need from the planet with no reference to possible limits of the capacity of the planet itself. That interpretation is unfortunate.

It seems fair to suppose that no such assumption resided in the minds of anyone on the Commission. The words that flesh out these ideas can be found elsewhere in the report's text. I spoke with Jim MacNeill, who served as Secretary General of the Commission and lead author of the document in question, *Our Common Future*. He made the point that the now-famous passage had not been intended as the sole definition. In fact he pointed out that the matters that concern us do not easily lend themselves to absolute definition. He expressed a degree of dissatisfaction with the fact that the particular extract was used as everyone's definition for sustainability or even sustainable development. The principal reason offered for his disappointment was that he felt that other parts of the report were routinely ignored. MacNeill pointed to other passages. He describes them in the Introduction to *Cents And Sustainability*. 'Development based on "consumption standards that are within the bounds of the ecologically possible and to which all can reasonably aspire" or development that, "at minimum … must not endanger the natural systems that support life on Earth: the atmosphere, the waters, the soils, and the living beings"' (World Commission on Environment and Development 1987), quoted in MacNeill (2010). He felt that the intense focus on what has come to be called the Brundtland definition of solely the intergenerational equity component perhaps takes a little away from the report's larger message.

Unfortunately, those particular words have become an iconic formulation of a common take on what sustainability is all about; many have based their views of sustainability entirely on these few bullet points, without expanding into the detail included in the rest of the report or appreciating the actual way that sustainability can become a quality of the operation of human existence.

Taken in isolation, none of this was out of place in the historical setting that prevailed when the Commission's task was set, thirty-odd years ago, but the way in which these few words were adapted as a definition can certainly now seem naïve or disingenuous. MacNeill notes that a good portion of the corporate world seems to actively resent any mention of limits and that this bias has worked its way into the broader public discourse. He recalled being invited to speak at a forestry conference and heard an industry leader proclaim the definition of sustainability that was in use. MacNeill found the definition hopelessly flawed and unsuitable, and when he asked where it came from, he was told that it was drafted at an industry gathering. That is the kind of error that the aviation and alternative fuel industries will have to avoid.

We should never infer from the Report of WCED (or anything else) that the environment is able to do anything that we ask of it; that we only need to organize ourselves better and deploy the necessary technologies into whatever mess we have made. We lose any sense of recognition of limits imposed by physical systems when we frame things that way.

In my view, the enthusiasm with which we have latched onto this passage seems to reinforce our species' disregard for environmental matters. We cannot quite believe that we will be unable to satisfy our needs, wants, or desires; we appear to assume that the planet has the capacity to provide the resources that we will consume in the effort—no matter what. In fact, technology *will* help, and so will adjustment of our political, social, and institutional framework. But only if we *first* accept the limits implied by a true comprehension of sustainability. The 'Brundtland definition' is just a phrase. It is entirely inadequate to the job of establishing and ordering our limits and, consequently, our priorities.

What everyone should perhaps acknowledge is that *Our Common Future* constituted

- an authoritative and complete statement of the concerns that the General Assembly recognized, and
- an expression of some of the views and approaches that would have to be part of the world's subsequent discussion and planning.

But if there is a 'theory' of sustainability that arises from that very long and fraught global discussion, it might well form itself around these simple ideas:

1. The environment, as the physical engine or plant that encompasses all of the world's processes, must be held sacrosanct.
2. Getting all the necessary people to participate in addressing our challenges (environmental, social, political, economic) requires that everyone be treated fairly and afforded equal opportunities.
3. Economic considerations need to assign actual economic value to the benefits and *costs* of the things that we do or make—both products and by-products.
4. For our global society to be sustainable, each of the things that we do or make must be sustainable in its own right, or at least working toward being sustainable quickly enough to avoid the worst effects that could occur before the transition is complete.
5. That there can be no disconnection between environmental, social, and economic considerations; they all interdepend absolutely.

We continue to feature the larger idea of sustainable development in terms of our aspirations for less-wealthy societies, but we now also think in terms of sustainability as a complex standalone criterion, a way of setting standards for anything that we might undertake. In the largest context, development is one element of global sustainability. Some may have used other reasoning to come to a positive view of sustainability, but it is probably enough for now that most of us agree on where we are, regardless of how we arrived. However, the fact that not everyone recognizes or accepts this new understanding creates an immensely important barrier to progress.

Seeing the Big Picture and Proceeding With Caution

It is undeniably easier to figure out what any individual initiative such as sustainable fuel production or sustainable food production would look like if we try to imagine it in terms of a larger (at this point, admittedly notional) goal of an entirely sustainable world. Because that encourages us to ask such questions as 'How does this fuel initiative fit within that larger context and those ways of doing things?' Such a comprehensive view allows us to diminish the effect of the phenomenon of creating a new problem by attacking too narrowly the one that preoccupies us. If we address an environmental matter here in a way that creates a social problem there, and turning to the social problem creates an economic roadblock that results in a different environmental problem somewhere else (and so on), we are not helping ourselves. The best approach is to move toward making each of our initiatives and each of the enabling actions within that initiative more and more sustainable in all dimensions and in their own rights according to precepts that we may discover elsewhere while trying to do other things. When we drive a car, the location of a distant point on the road, visible on the horizon, has no particular relevance to our task. On the other hand, the distance of the car from each edge of our lane and the precise direction that the vehicle is pointed do. Yet we do not drive by preoccupying ourselves with these latter particular things so much. We look down the road and by paying attention to a more distant target, we learn that we will naturally do those things that will keep us straight and on the right path. Our hands cannot manage our direction and location precisely unless we keep the more distant aim.

The Perfect as Enemy of the Good

We have considered over and over in these pages that action has effect. Since virtually all effects occur in degrees, and since we know that our very existence creates *some* effect, the first danger is in regarding sustainability as something that can only be framed in absolute terms (a practice is either sustainable or *not* sustainable, full stop). Not only do we miss the point, we miss the crucial opportunity to assess and adjust according to our understanding. We cannot create a way of living on Earth that allows us to claim that we're having no effect at all. Requiring perfect solutions all at once really becomes an excuse for doing nothing. We will surely change things, and we need to do so cautiously, but we do need to act.

The conception and development of a theory of sustainable action, as opposed to a 'definition' of 'Sustainability', allows us to gauge the relative urgency and need to adjust action (advancement, development, exploitation, increase or reduction in magnitude of effect or presence) with moderation and control of that action. And it is a postulation that allows for the comprehension of our capacity for doing so. It is an idea that has all the power in the world to allow us to succeed or (should we fail to internalize it) to damage our world, its supporting ecosystems, and ourselves, slightly, horribly, or to any degree between.

What can be done varies greatly from place to place and even from interest group to interest group. In his book *Sustainability: a Philosophy of Adaptive Ecosystem Management*, Bryan Norton summarizes that theorists have polarized debate by seeing issues only through their own academic specialist lenses, whereas public discourse takes a pluralistic view of environmental values that allows for compromises and balances (Norton 2005). In going beyond the environmental, we have to accommodate pluralistic views but in more spheres, and in a way that allows us to ratchet up progress in one area while simultaneously paying attention to the others. This would argue that when we talk about efforts to assess or certify an object of interest (fuel, for example) with respect to sustainability, it is wise to make reference to both its degree of compliance with *all* accepted criteria of sustainability, and also to its apparent *relative* degree of sustainability (as opposed to an absolute sustainable/unsustainable set of determinations).

But it is problematic to swap factors that we cannot really quantify or compare properly, even if we could quantify them—which is almost always the case in this discussion. Certainly we cannot jump out of the starting gate with zero-carbon fuel, so it seems obvious that we will be assessing relative advantage there for a long time. It is equally obvious that it is just not acceptable to (for example) disregard matters of social justice here to afford 'more' social justice somewhere else. We will talk more about this in the following chapters, where we deal with assessment of sustainability and then policy. However, we should perhaps remember that the most prominent reason that the world is engaged on this problem at all is that we have been swapping improvement in wealth, in certain parts of the world and for certain people, for loss of a particular environmental capacity, at a global scale, and whose effects are unevenly distributed.

Putting Sustainability First

We are about to transition to our discussion of sustainability assessment. An article that is useful in showing that there are great challenges in coming to common understanding of sustainability and its assessment is Pope, Annandale, and Morrison-Saunders' 'Conceptualising sustainability assessment' (2004). One significant observation that these authors make is that the right time to think about what is being undertaken is *before* one starts; we are prone to devising plans, working them out in detail in terms of their cost and revenue, and *then* seeing whether they can be done sustainably. And sustainability initiatives (sustainable fuel would be one) would be facilitated if we adopted the habit of considering beforehand what sustainability demands: it is easier to view projects through a sustainability lens while planning than to attempt to make them be sustainable once undertaken. Similarly, anyone who is interested in gaining an understanding of sustainability or developing policy in that field should have a pretty good idea of what sustainability means in the larger public forum rather than wasting time and money in making post-hoc adjustments to poorly conceived initiatives. Let us think hard about what has been said about sustainability before we advance an idea that we expect the world to accept.

This becomes an exercise in understanding interest, making selections, and internal editing—coincidentally, one could hardly find a better definition for 'political thought'. Political exigency indeed provides context, and we need to realize what is at play by hearing different voices. Sustainability will be accomplished by degrees, in different ways, in different places, and I do not want to suggest that once we understand it, we can just flip a mental switch and 'do' it. But if sustainability is an ideal that we move *toward*, we need to have absolutely the best idea about what we are trying to do and where we are trying to go, if the pursuit is not to result in cost, loss, waste, time, and frustration.

The caution about thinking first and acting incrementally or progressively applies to the process of developing standards and criteria just as much as to the care with which we vet actual physical projects. We must get standards right.

The Air Industry as Poster Child for Sustainability

Not every country, industry, or individual will see the benefit of a comprehensive conceptualization of sustainability accompanied by a will to act on this understanding. The air industry is first among those who do see it, a perspective based in part upon the absolutely astounding commitments in time and capital made routinely by governments, manufacturers, airlines and air industry suppliers. Of all of the world's activities, aviation is the one that is most globalized by its very definition. And of all of the activities where long-term investment in capital assets is key, few would top it.

Aviation is at a 'tripping' point, a point of vulnerability as it commits to sustainable fuel. Extreme care must be taken to establish some internally consistent industry meaning for sustainable fuel that is in accord with each of those meanings that have established a beachhead in parts of the public consciousness. I am of the opinion that changing to sustainable fuels is worthy on its own merits, but for others (and, realistically, there are hugely important commercial factors at play here), a large part of the value in proceeding toward sustainable fuel is to win a critical public relations battle for commercial air travel. If winning that battle is important, the only way to win it is convincingly, by knowing and addressing everything that appears as a sustainability factor in popular discourse, and by proactively assessing potential solutions. It should become the goal of the air travel industry to be so thorough and committed to securing sustainable sources of flight energy that it will set rather than match pace; that it will lead rather than follow; and that its standards will be praised rather than needing to be defended.

Commercial flight should make itself a showcase for the kind of thinking and practice that advocates of sustainability would want to support.

In the alternative, it is certainly tempting to hedge. Aviation, like some other industries, could pursue a policy of doing less while trying to make its limited efforts seem like a bigger commitment. The air travel industry—manufacturers, airlines, air traffic managers, regulators—has already done a great deal (a lot more than many others) in reducing unit fuel consumption, so a defensive attitude would not be surprising. Though justified and normal in most similar circumstances, it would, however, constitute a kind of failure for the industry to find itself disputing accusations that it is failing to be sustainable, or making apologies for performance that may be limited by real obstacles. A cynical public, roused by very concerned and skeptical opinion leaders, will pick apart every claim, reveal every bit of hype that is mere greenwashing and expose exactly what they think air carriers and others are really doing—or not doing. And there are certainly enough cynics to make a difference. Activities like flying, which represent extremely concentrated use of energy and production of GHGs, will be subjected to the highest scrutiny, especially when examined by those who have little opportunity to benefit from such activities. In this kind of battle for the public mind (already misled by different industries advocating in their own interests) the only way to win is overwhelmingly.

Chapter 9
Measuring Sustainability

Getting Real

In the previous chapters, we talked about the nature of sustainability. In this one we will talk about how we make determinations as to the sustainability of aviation and the initiatives that aviation undertakes to improve its sustainability. Most notably, we want to know how we would assess the sustainability of the processes that would provide new sources of flight energy, designed to reduce aviation's carbon burden.

When we examine what is being done to try and assess sustainability of materials, we will realize the relevance of the more abstract discussion that has preceded. Things that may have seemed abstruse or only tangentially related to matters of 'real' sustainability will appear now as the stuff of hard or at least authentic judgment.

Very early on, the point was made that people, particularly those who hold sustainability as a priority themselves, will watch the process very carefully to see just how valid the sustainability claims about new fuel might be. One thinks of environmental non-governmental organizations (NGOs). Certainly they will be looking for the direct connection between the theoretical consideration and the physical fact. We just heard from Jim MacNeill on the subject of sustainability. He was interested to learn about aviation's sustainable fuel efforts, but he was also cautious. I had explained some of the airline and alternative fuel industries' goals, and in describing some of the certification efforts, about which you will read here, I claimed that they were being very conservative. On that point, he offered the following comment: 'Well, they had better be. People who run very large, well-financed environmental organizations with large research arms will be looking at those claims very carefully. Any suspicion that they derive from the need for a good public image rather than from actual achievements will be disastrous for the industry.' In my research, I did not meet anyone in the airline business who would disagree with that.

What is Sustainable Fuel?

Principles of legality are not easily married with principles of sustainability. There are a number of reasons for the difficulty with formalizing provisions of sustainability, but two challenges are:

1. many sustainability factors are hard to measure at all, and
2. we remain a certain distance away from perfect measures of things that *should* be measurable.

We can see some of this effect when we watch different jurisdictions try to accomplish a generally similar goal, and a problem arises when different entities simultaneously embark on parallel and unlinked efforts. Since the whole project of assessing and certifying sustainability is economically challenging and politically fraught, even well-intentioned nations can very easily be disheartened and withdraw their support if they see that each country's expensive and time-consuming work produces different results, even in minor details. Everything is important at the margin and minor details are major for someone, somewhere, you can be sure.

Consider how many separate jurisdictions might be interested in enacting a sustainability initiative (alternative fuel would be only one example). If we think about fuel, there are quite literally thousands of entities that could aspire to establish a standard. Each nation, state, province, territory, or municipality might reasonably want to do this; it is an important subject and lots of us want to get going. But it would be useful if international airlines and their home governments would support a single certifying mechanism rather than each trying to tailor a standard to their own needs. Imagine, for example, how difficult it would be to convince the flying public (or anyone else) that commercial aviation was committed to sustainable fuel if sustainability was always being defined, measured and certified differently, depending on the policy of the government where the fuel was boarded. And individual and solitary efforts are the least credible; when one entity uses its own criteria, there is no assurance in the public eye that it is not somehow less ambitious and disciplined. In the event, someone's own criteria presumably *would* be lax. Why else have different ones?

Legal restrictions or standards should logically use values that have already been developed either collectively or through the offices of credible and unbiased authorities. Unification of a standard (if a strong one is agreed) takes special interests out of the equation. That is important. Standards that are advanced by a government with a particular interest or an organization whose members have a strong incentive toward an advocacy role, cannot be relied upon as confidently. For example, the voluntary standards developed by the Council on Sustainable Biomass Production (CSBP) for the sustainable production of agricultural biomass in the US (Council on Sustainable Biomass Production 2012) {Council on Sustainable Biomass Production, 2012 #134} are laudable, but do they lack the rigor and comprehensiveness that come with efforts that are driven by a broader constituency? Vested entities such as states and industry advocacy groups can be relied upon to run an organization that administers a standard (though that is not without hazard) but efforts to *develop* such a standard should be collaborative and proceed with the involvement of all interests.

Cramer Commission

In the Netherlands, there was a desire to ensure that biofuel production would constitute a step toward greater sustainability. A Dutch assessment protocol for sustainability of the production of biomass was sought, and a commission to address this was established, chaired by Jacqueline Cramer. The commission produced a very useful report (Cramer et al. 2006), as well as a follow-up report on a testing framework for sustainable biomass (Cramer et al. 2007), which make quite clear that standards generation cannot realistically be left to corporations. That one finding is important and still valid, especially for the air industries. Appropriately, the Commission also suggested that the Netherlands government should engage other policy-making groups at the European and international levels in the development of a perfectly common or equivalent standard, so that there would be no debate about the meaning of commitment to sustainable biomass. The efforts of the Netherlands' Cramer Commission and the UK's Renewable Transport Fuel Obligation[1] gave rise to a larger initiative, the European Union's Renewable Energy Directive (EU RED) (European Parliament and Council 2009a). Much of the Cramer Commission's reports can be seen reflected in the standards promoted by the Roundtable on Sustainable Biomaterials (RSB),[2] which will be discussed later in the chapter. The point is that evolution toward higher and more common standards is the essential path.

Aviation Makes its Own Argument

There is a little booklet that describes the need for commercial aviation to change how it sources fuel. It says that sustainable fuels are crucial, and that biofuels can fill the bill. It talks about the possibility of seeing 80–84 percent reductions in CO_2 emissions on a life cycle basis. It claims that in the past, biofuels have competed for land and water with food crops but contends that this should not happen any more. It describes attributes of suitable, sustainable biofuels:

- cheap;
- made from plants that do not take up arable land;
- do not employ excessive farming techniques, such as:
  – excessive amounts of pesticide,
  – excessive amounts of fertilizer,
  – excessive irrigation;
- do not threaten biodiversity;
- provide socio-economic value to local communities;
- produce lower amounts of carbon on a carbon LCA basis than conventional sources of jet fuel.

---

1   See https://www.gov.uk/renewable-transport-fuels-obligation
2   See http://rsb.org/sustainability/rsb-sustainability-standards/

And it says that all of these things constitute only part of what would be required of sustainable biofuel candidates.

Except for the fact that low, stable price is featured prominently as an essential quality, it would be reasonable to assume that the foregoing was drafted by a moderate environmental organization. In fact, it is an industry position. In this booklet, *Beginner's Guide to Aviation Biofuels* (Air Transport Action Group 2009), the Air Transport Action Group (ATAG) describes the air industry's commitment to sustainability in fuel. ATAG represents a broad cross-section of air industry players and is certainly credible in expressing commercial aviation's position. Prominent voices in the air industry community want a high standard and they insist upon evolution toward a *common* standard because airplanes fly everywhere. What good is a standard that is lower than another standard where an airline operates? Everywhere an airplane goes, it arrives with a quantity of fuel that was bought somewhere else.

No matter what one feels about aviation's actual chances of achieving these economic, environmental, and social justice goals, and accepting that the description of 'sustainable' biofuels might provoke a quibble or two (or even more than a quibble), the ATAG statement is, nevertheless, evidence of a remarkable level of commitment to the idea of sustainability—and sustainability as a multidimensional quality embracing the very elements that we discussed in the previous chapters. While it might gloss over a few points, the booklet does mention that in order for the airlines to be sure that their biofuel came from 'truly sustainable sources', important work in *certifying* such fuel as truly sustainable would have to be undertaken, and mentions the work of the Roundtable on Sustainable Biofuels (now the Roundtable on Sustainable Biomaterials, or RSB) and other very credible initiatives that are setting about ways of examining and certifying fuel production on the basis of elaborate sustainability criteria. Some critics claim that the sustainability goals are not perfect, while others predict that aviation will not meet them. That is hardly the point. The point is that making a commitment and getting started are all that are required and the brochure sets up the process quite well. Let us see how well it conforms with the work being done by multi-stakeholder efforts to enact a calculus of sustainability.

The Scope of Sustainability Assessment

In reality, sustainable fuel has already started production at this writing. And it is being certified. So this discussion is not all in the future. But we are still learning about what can realistically be included under the rubric of sustainability, and need to determine what the criteria should look like as they evolve. What things should be captured by a proper standard? In the following sections we will examine what is probably a fair selection of the things that are currently considered to form part of the sustainability equation—physical and social—with the understanding that this treatment is necessarily cursory.

Air

A dominant theme of this book is the matter of our effect on the Earth's atmosphere. There has been a great public discussion about how raising the amount of GHG in the atmosphere causes global warming, which affects patterns of air circulation and the hydrological cycles so that local and regional climate drifts away from norms, and the severity or frequency of extreme weather events, such as storm, drought, and flood, increases. These changes perturb habitat for all plant and animal species; those whose habitats become insupportable may ultimately decline, thus affecting other species and further altering local ecologies. These compounding climatic and consequent ecological changes will have a profound impact on human populations when they affect food production, water availability, and demands for energy and materials.

Clearly, if air travel's flight energy requirement is to be satisfied in a more sustainable way, we will have to move from fuels that increase the net amount of CO_2 in the global atmosphere to fuels that do not—or at least not as much. Any industry that sets out to produce a fuel—whether traditional or 'new'—must account for its own production of GHG; analysis of the carbon content of aircraft exhaust thus requires a capacity to track carbon uptake and emission though the entire fuel production and use cycle.

But all of the non-GHG matters pertaining to contamination of the air also apply. The activity of flying and the processes that might be used to produce a candidate fuel will always come with a broad set of implications, of which local air quality is one. What we will measure in terms of the air will be the carbon balance of the fuel that is produced (to what extent its use results in lower carbon emissions when the carbon *captured* during its production is netted out) together with any local effects on air quality where that alternate fuel is produced.

Water

Similarly, the direct effect of GHG and global warming on ocean systems and the hydrological cycle is staggering. As the CO_2 that preoccupies us is brought to greater levels in the atmosphere, it necessarily finds ways to spread itself around. The ocean acts as a huge sink for CO_2—but not without limitation or consequence. Ocean water becomes more acidic, with an enormous impact on sea life. And the global warming of the atmosphere affects the aquatic environment as well: changes in water temperature affect sea life directly and also disrupt patterns of circulation, creating even more profound changes in all the various localities.

But again, production of novel fuels can have its own effects of GHG production, contamination, and overuse of resources. To be sustainable, processes such as biomass cultivation and fuel manufacture must not deplete or contaminate local fresh water supplies including ground water, and must maintain adequate flows of river water so that downstream environments also remain viable. They must not disrupt circulation or change temperature in a harmful way, nor contaminate local water such that an ecology is compromised.

Minerals

Even in the very earliest stages of development, humans extracted and discarded non-renewable mineral resources. While unsustainable in a theoretical sense, such small-scale use of available deposits for the immediate needs of small local populations constituted an inconsequential depletion of those resources. However, until a change of our total energy and materials regimes allows for the synthesis of useful materials from renewable energy stocks and pedestrian materials such as common rocks, or gases, or perhaps even ocean water, it is now considered essential to preserve non-renewable mineral resources under terms of sustainability. We can gradually reduce our mining through efforts such as better recycling, for example, so that we do not deplete valuable minerals before they can be substituted for in some other way. In any event, any fuel making a sustainability claim cannot depend on the exhaustion of non-renewable mineral resources.

Land and Land Use

Once again, climate change has its own set of effects. Loss of the use of land through drought, change in sea level, erosion, and other climate effects is a threat, but production of a novel fuel also has potential effects that are profound and among the most controversial when studying the viability of alternative sources of fuel. Most obviously, land should not be degraded, polluted, or lost (through erosion or subsidence) as a result of fuel production. Equally important is the question of whether land's use is altered, and thereby compromises the services that it formerly provided. Mature forests, grasslands, and wetlands contribute enormously valuable functions from an ecological and environmental point of view, serving as water stores, climate moderators, and a sort of air conditioning system that helps to maintain local air purity, as well as water quality and availability. Of course they also provide habitat, often including the complex combination of setting and resource for human society. Critically, they also contain enormous stores of carbon in the planet's carbon cycle; destroying these ecosystems for an altered land use releases that store of carbon into circulation in the atmosphere as a current excess, and also reduces the land's capacity to store carbon in the future. Taking forest and converting it to produce biomass for fuel is an example of direct land use change (LUC), but even if we use land that is neither forest, nor prairie, nor swamp, we can still bring about devastating effects.

Consider the case of agricultural land that is used for fuel biomass production rather than for its original food or fiber crop. The use of the land has not been directly altered—it was growing an agricultural product before and that is exactly what it is doing now. But if the crop is changed from a food grain to a fuel biomass, it may affect the demand for that food grain: the commodity price might rise or persist at a higher level, when other factors might otherwise have led to a drop. This can result in other acreage that was in forest or grassland being stripped of its natural cover and converted to agricultural production of food to compensate for the loss of land that has gone to fuel production. In this way, a mere change of crop *here* induces a change of land use *there*. This is an example of indirect land use change (ILUC). It has proved to be extremely difficult to assess indirect land use change, because many other factors also affect food commodity price and availability. Higher prices obviously mean a marginal reduction in food accessibility for *someone*, but the cause may be hard to determine. Nevertheless, it is clear that we cannot take arable land out of food production and not expect to cause a downstream land use consequence somewhere.

Manufacturers of fairly conventional biofuels such as corn ethanol may take the position that they should not have to change their practices until someone is able to explain in detail and quantify accurately any negative land use consequence of their activities. Realistically though, who accepts this kind of defense? Perhaps it is not quite so important now, at the incipient stages of a move toward more sustainable fuel, where near-term targets are for modest levels of carbon reduction. We could almost say that if we had a notional goal to reduce fuel carbon by 1 percent, everything we measure would be in the 'noise band'. But this argument will not hold as more aggressive targets are established for years and decades out, and aviation will need to have the ILUC factor well in hand if it is making a long-term commitment to sustainable fuel.

Life-cycle Analysis: Understanding Material and Energy Flows

When we look at the threats to the physical environment, we realize a need to measure and quantify. Life-cycle analysis or assessment (LCA) of a product or manufactured commodity is the practice of creating an inventory of flows of materials and energy into and out of its production. Using such methods, we can collect and organize data on fuel production in such a way that we see whether certain production methods or facilities cause an increase of harmful by-products or deplete stores of important material resources. We can assess whether the energy used in the production and distribution of the fuel is itself a renewable form of energy and examine its components and carbon content. These observations can allow us to see the flows of energy and material all the way from source through production of our product and back again to source (if that is occurring). Variations on LCA (also called 'cradle to grave' or 'wells to wheels' assessment) help us understand the true environmental cost of production. A well-elaborated LCA is invaluable in coming to terms with these costs and the consequent real net benefit (or otherwise) of employing the product.

For example, corn ethanol is a fuel idea that is preyed upon routinely by environmental advocates, and whose GHG balance has been the subject of hot criticism and studies. It is not useful here to take a position on the sustainability worth of corn ethanol as a biofuel, but it *is* useful to understand what all of the fuss is about and how proper LCA can help resolve matters. Corn-sourced ethanol, requiring massive inputs of fossil fuel energy for its agricultural production and post-fermentation distillation, together with the burning of natural gas to produce its required fertilizers, seems to be not nearly as environmentally progressive as first thought, and has, understandably, provoked great argument about its use in advancing the cause of sustainability. (Searchinger et al. 2008). But it was the employment of LCA tools that allowed us to see any of the weaknesses in commercial corn ethanol's capacity for reducing atmospheric carbon, for example. All of the CO_2 that is released when we burn corn ethanol may well come from the renewable corn crop, but the CO_2 that is released by the burning of fossil fuel in trucks and tractors and other equipment that supports the growing of that crop also needs to be accounted for.

Such analysis on any particular criterion is an extremely challenging test of our ability to perform measurements and integrate information, but without it, we are nowhere. International, interjurisdictional, and intersectoral agreement on terms for LCA will be one of the most important elements of what we undertake when we set out to make human activity more sustainable. And in my view, the aviation industry should insist on the highest standards and measures of sustainability including exhaustive LCA. Where those standards fall short, aviation should be aggressive in demanding more from the certification establishment.

Genetic Risk

Many of the fuel technologies that will develop over the coming years will involve the use of genetically modified organisms (GMO). Current technologies, referred to as genetic engineering (GE) and synthetic biology (Synth-bio) are developing rapidly. Synth-bio techniques actually use genetic building blocks to customize an organism by having it incorporate new characteristics. The tailored genetic design of feedstock material or perhaps microbial organisms employed in converting even non-biomass feedstock to fuel will be accomplished using these powerful technologies. In cases where novel life forms are created, there has to be absolute assurance that new genetic traits cannot expand into the broader environment outside of the production area. Propagation of the new organism or hybridization with wild species in the open environment is a process that has the possibility of running absolutely unchecked, constituting a potential ecological threat. Some past assurances by the producers of GMOs used in agriculture have not held, and many individuals, interest groups, non-governmental and governmental organizations have weighed in on the topic of 'genetic pollution'. (Ellstrand (2003) provides a scientifically sound, reasoned overview.) So any claims that risks of uncontrolled propagation of a species or trait in the wild are extremely low will not be accepted by large segments of society that are wary of these technologies. If genetically engineered or synthetically bio-engineered species are employed, it must be on the basis that the chance of proliferation is essentially nil—a very high bar. Even then, it will be a tough sell. Some potential fuel producers rely on the exploitation of novel life forms in closed bio-reactors, while others see more open areas of growth. Standards, if they are credible, will insist that bio-tech advocates demonstrate zero risk of danger to genetic stability in wild populations of life.

Biodiversity and Ecologies

The way in which materials and life interact on the planet's surface is such a complex evolution that there is no real way to describe it in detail, but recognition of the need to allow for the continuation of both broad and local ecologies is not merely a politically correct sentiment but a key piece of the sustainability puzzle. Altering the interdependence of species can alter the whole global ecological structure, which is critical to us all. Even when human action disrupts *merely* a local ecology, it can have profound effects on the ability, for instance, of people in that locality to provide for themselves and their descendants, thus creating economic and social justice consequences that make such action entirely unsustainable.

Social Justice

The social justice aspect of this discussion is enormous. Does our fuel production introduce significant imbalances in the sociocultural framework of the jurisdiction where the production is located? Are workers fairly treated? Is land left in the hands of those who traditionally exploited it? Does the insertion of this new enterprise into a locale create an inappropriate disruption as regards governance, politics or policy? How do we know? Is the area governed in a transparent way that lets us see how our actions will affect the people? Is there anything that we would recognize as effective government at all? Are there other ways of knowing that the people most directly involved understand and support the enterprise?

Note too that all of the issues related to the planet's physical assets (air, water, minerals, land, and so on) have a clear bearing on matters of social justice. Loss of a key water resource changes life for a community, and even the hardest-to-assess aspects of the physical world bear on justice. For example, land tenure that is affected by efforts to secure land for fuel production is already an enormous issue. A 2012 International Land Coalition report indicates that between 2000 and 2010, approximately 71 million hectares (mHa) of land in rural communities have been acquired by commercial interests, often foreign; although much of this acquisition is spurred by concerns that have been heightened during recent food crises, the largest single driver of this land rush is biofuels. Of the 64.3 mHa of acquired land whose purpose is known, 37.2 mHa—58 percent—are devoted to fuel production! This is more than three times the amount of land that was acquired for food production (11.3 mHa) and more than four times the amount for forestry (8.2 mHa), the next two largest acquisition purposes. The Coalition report outlines the global commercial pressure on land and points out some of the obvious vulnerabilities of rural and small landholders whose land tenure may be compromised to their own specific detriment but upon whom additional billions of people in the developing world depend for food (Anseeuw et al. 2012).

Land use is clearly a sustainability issue on many fronts. Concerns about appropriation leading to loss of ecologically and environmentally valuable woodland or grassland, for example, should be coupled with social justice and economic consequences for the people who are using or depending on that land. Access to water and other resources, preservation of cultural patrimony, and traditional values are all things to which considerations of social justice must pertain.

The countervailing promises of jobs with better incomes, export opportunities, more reliable access to better food, and other benefits arising from these promised jobs and restructured communities must be accompanied by completely dependable, honest information and truly informed consent that thoroughly comprehends risks and benefits. Decisions from above, made by governments or developers, are not appropriate in the consideration of social justice. Being advantageous to the most people or the average person or *some* people is not the same as being fair to *all* people. The decisions to engage themselves, their families, their communities, their land, their resources and their ways of life are decisions that the affected people themselves must own entirely. Whether or not local government can provide adequate mechanisms for engagement, such as forms of participatory democracy, certification must ensure that informed consent is driving all decisions.

Development Goals

Because economic and social justice dimensions must be accommodated in delivering more sustainable services to our world, and recognizing that some developing countries may act against costly efforts to achieve sustainability unless such efforts can overcome a currently hopeless future for themselves, we can now see and address the matter of legitimate development aspirations in the context of trying to produce better fuels. In indirect ways—though not likely to be part of sustainability assessment criteria—locating new fuel enterprises in the developing world may form a part of sustainability strategies in the best interests of all. However, though fuel fungibility argues that fuel can be made and boarded anywhere, regardless of who the user is, there will be obvious sustainability benefits in producing fuel nearest to where it is loaded on aircraft. As we have discussed, current fuel use overwhelmingly occurs within the developed or rapidly developing worlds, and that scale of demand may swamp the ability of smaller centers to produce and integrate new fuels. Also, the developed world will see security of supply benefits through achieving local production. But developing countries must at least be given the opportunity to produce local supplies if that is possible, and if fuel can be as sustainably produced in the developing world as the developed world (all transportation and other matters considered) then perhaps fuel production can facilitate technology transfer and development.

A bridge comment here that will take us into the next topic is that there has been a preoccupation with scaling up potential fuel production technologies so that they make commercial sense. In deploying these systems into the developing world, there may be challenges in scaling them *down* so that they operate at volumes that are appropriate to smaller local demand.

Economics, Including Externalities

A system that is economically nonsensical or inoperable will not be adopted, so the production of sustainable fuel must be economically viable. Obviously, for commercial flight to move toward greater degrees of sustainability, the production of alternative fuels must make economic sense at some level and at the largest scale in order for our societies to act, either collectively or individually, to create those fuels.

That does not mean that it must be immediately profitable without aid or subsidy, only that an appropriate amount of value is created for those incurring the costs. So even if some fuel production is not commercially worthwhile at the outset, if governments see economic value in fostering its availability, it may be considered economically sustainable in the context of that larger set of considerations, and especially in the early going. This has important implications from the point of view of policy-makers. Since concentrated capital is accompanied by powerful capacity to advocate, governments and the public can often be swayed to accept particular perspectives on economic viability as defined by profit, jobs, and tax revenues. These are very important things. But the sustainability discussion has to take the longest and largest view in order to generate the greatest benefit, rather than a benefit that suits shorter terms and narrower concerns. It is in the best interests of both the air and alternative fuel industries to help the public understand economic viability and advancement, and how sustainable fuel can satisfy these criteria, because voices from other particular industries or constituencies will offer countering views on all of this. It is not necessary that existing corporations make money manufacturing sustainable fuels, it is only necessary that eventually *someone* make money.

Even if there is a general acknowledgement of the need to ensure that specific goals, techniques and projects are economically sound, we must also insist that those who are charged with such analyses include, as an essential component of that work, a full accounting of all externalities—the unintended economic impacts (positive or negative) of activities or transactions carried out by unconnected parties. While there is no point in supporting a solution that actually creates or increases problems, there is also no point in dismissing an initiative or strategy without examining its potential to address some externality that might otherwise constitute a large cost. In other words, techniques like LCA are more easily applied if we quantify the economic harm of unconsidered externalities rather than merely contending that they are bad. Understanding the full range of flows of economic effects in societies and in the lives of corporations and individuals is important, so a vigorous effort to inventory and quantify externalities constitutes a de facto prerequisite to sustainability assessment.

Here the intergenerational aspects of sustainability are key; before dismissing an initiative as unsustainable on an economic basis, we must consider the benefits over time. Much of the debate about proper discounting of near-term capital investment would perhaps be resolved if better efforts were made to estimate and evaluate the progressively expanding long-term effects of those things that have, thus far, been ignored as external to our economy.

My Sustainability Versus Your Sustainability

We are accustomed to hearing about economic sustainability or unsustainability as assessed in terms of the relatively immediate needs of individuals, corporations, towns and other players operating at a very limited scale. But sustainability is not just about the persistence of an entity, it is about persistence of the whole; assessment and balance are essential at every scale.

On the environmental front, much is dismissed as economically unsustainable on the basis that it costs too much to be profitable or affordable in the nearer term. This is actually a question of political supportability, not sustainability. Ignorance and insincerity may lead to an initiative being criticized as economically unsustainable: we either do not really understand the consequences of failure to implement it or simply put our own immediate interests first, and hope that someone else will take up the slack. The financial burden that associates itself with action is not economic unsustainability, it is cost. As long as benefits accrue and can be reconciled, we meet the criterion of economic sustainability. It is not necessary for the market to clear a value that results in a benefit for specific people, or for our immediate and local purposes.

We can see from the foregoing that assessing sustainability and measuring all of the elements of which it is comprised will be a challenge. In fact, it will be a new way of doing business, commerce, and trade. As soon as sustainability gains traction with aviation fuel, it will expand upward into those materials and energy sources that supply into fuel production. If production of those things must become sustainable, the concept penetrates more deeply into supply chains. Who will develop ways and means of assessing sustainability so that their approach can be cloned by more and more supplier enterprises? In the next chapter we will learn more about that task and a particular enterprise that is prepared to take it on.

Chapter 10
Sustainability Assessment and Certification: Who Should Do It?

The job of assessing sustainability will be tough, but there are forces that will not even want us to try. Commercial interests whose actions tend to render the idea of a sustainability standard meaningless by making it undemanding and marginal in terms of thoroughness, comprehensiveness, and broad applicability will ruin the commercial prospects of those who seek to provide sustainable services and goods. And the situation is precarious because there is lots of inherent pressure in that direction; activities that are not sustainable see sustainability as a threat. In that context, sustainability 'camouflage' in the form of a lax standard allows for some equivocation that is useful to those who seek to do nothing and to be left alone.

Establishing uniform sustainability measurement is unlike other endeavors. For example, if we move away from sustainability for a moment and think about the standards system that we use in determining the *technical suitability* of a fuel for the aircraft and engine, approving fuel for use where there is worldwide recognition of tolerable limits for being safely stored, handled, and burned in is relatively straightforward: All certification organizations work with specific limits on specific criteria, and therefore no real economic or commercial advantage exists for choosing one organization's certification over another's. The type of fuel that must be burned is spelled out in the aircraft operating manual *Limitations* chapter; a technical standards body that did not measure to those specifications just would not be of any use.

That is most assuredly not the same as assessing sustainability. If a 'sustainable' fuel is preferred or mandated, and if an airplane can burn a fuel safely regardless of why or how it is determined to be sustainable, there are all kinds of incentives for certifying a fuel as sustainable on the basis of lowered costs or some other advantage to the entity that wants to host and control the certification standard and process. At this early stage, all of the individual components that make up the dimensions of sustainability can be seen as arbitrary in nature and degree. And assessing them, where materials, methods, costs, and interests of states, producers, and users vary at every turn, is clearly problematic. A little later on in this chapter, we will describe what I think is the most advanced and robust effort to certify alternative fuels as sustainable. In the meantime, let's look at the goal.

Assessment Challenges

It is a patent understatement to say that accomplishing meaningful and comprehensive evaluation of the sustainability of *anything* will be difficult. Given all the things that we have had to address in understanding the concept, the prospect of having to measure it is obviously quite daunting. But just as a teenager is perhaps staggered by the idea of having to clean up their room, it will behoove us to remember the importance of simply picking up that first sock. And thanks are due to those brave enough to have started us on the road to sustainability by actually setting out to describe its criteria in concrete terms and establishing means of assessing things such as fuel in those same terms.

One particular issue that complicates the task of assessment is attribution—determining who is responsible for what. For example, deciding if a fuel is sustainable depends on two factors: whether the process of making the fuel is sustainable, and whether using it renders the consuming activity (flying in our case) more sustainable. So the question of attribution of (say) carbon production in both manufacturing and using the fuel is absolutely critical. In fact, when we say that the industry aspires to eventually secure sources of nearly 'carbon-neutral' fuel, that is a bit misleading. We really want to imply, in a way, 'carbon-negative' fuel; making the fuel needs to take at least as much carbon out of the atmosphere as the airplane is going to release in burning it.

Where Does the Assessment Occur?

This uncovers a larger topic that is fraught with political and policy peril, operates within and between industries and in all jurisdictional contexts, and lurks at the edges of every sustainability discussion: *where* and *in whose hands* will the sustainability effects occur and be accounted for? Understandably, entities prefer to favor their own interests, so in general, producers tend to think that consumers should take responsibility for consequences. For example, poorer countries that manufacture consumer goods for markets in wealthier countries often feel that those wealthier countries should be answerable for the carbon released in manufacturing the goods. But the consuming states argue that they have nothing to do with the release of the carbon, since the goods are made far away from where they are used; lower costs attracted that manufacturing activity to the producer nations, who should factor costs of achieving sustainability effects against their economic gains. The consuming nations have a point: otherwise, how do we control for 'leakage', where capital exits an economy rather than remaining within it? Suppose Country A enacts a strict environmental law that is intended to reduce the production of carbon in the manufacture of, say, furniture; this may create a cost for Country A's furniture industry. Country B, on the other hand, disregards carbon emissions from furniture manufacture, incurs no such cost, thus gains a commercial advantage, and exports its furniture products back to Country A. Not surprisingly, Country A's furniture industry might object and resist. In order to preserve the effect of its carbon initiative and protect its economic interests, Country A must restructure its commerce and trade balance, consider trade sanctions, or face other losses of effort and economic activity. Why would a country give up an important industrial sector for environmental reasons, and then see that progress wiped out by other countries that are willing to ignore those reasons and even increase their profits due to the first county's conscientiousness?

However, in the context of a world economy where countries seek access to each other's markets, trade sanctions are frowned upon, as are barriers to bilateral and multilateral trade. Unless furniture manufacturing is critical to Country A's economy, its best option is often just to grin and bear it, accept to trade in the offending goods and hope that the carbon content issues are resolved in future agreements.

Sometimes, if there is concern, an exporting (B) country might wish to retain responsibility for carbon as preferable to losing the opportunity to export. They will not want the people in the client nation to feel the carbon burden as disincentive to consume. They will wish to keep responsibility (perhaps especially if their intent is to somehow avoid penalty).

But in certain particular circumstances, consumer (A) countries will certainly be insistent. Energy products are special, and domestic production of energy (or even the possibility of domestic production) is always a matter of importance. For example—and still related to fuel—the European Union (EU) drafted an amendment to its Fuel Quality Directive,[1] whereby fuel imported to the EU *includes* the implicit life-cycle analysis (LCA) carbon burden associated with its manufacture. The Directive's Article 7a (new) includes the following provisions:

> With effect from 1 January 2011, suppliers shall report annually, to the authority designated by the Member State, on the greenhouse gas intensity of fuel and energy supplied within each Member State by providing, as a minimum, the following information:
>
> a. the total volume of each type of fuel or energy supplied, indicating where purchased and its origin; and
> b. life cycle greenhouse gas emissions per unit of energy.
>
> Member States shall ensure that reports are subject to verification. (European Parliament and Council 2009b).

Under regulations such as these, the carbon burden of all fuel supplies is measured and reported; countries can then make a clear case for policy decisions based on the carbon burden of fuels *now in their hands*, whether those fuels are imported or

---

1   See European Parliament and Council (1998) for the original Directive, and also http://ec.europa.eu/environment/air/transport/fuel.htm for an overview of subsequent amendments

developed domestically. This allows the EU to set low-carbon standards on fuel and to develop low-carbon fuels in competition with those they are importing. This is a direct encroachment on the exporting country's otherwise normal right to say, 'Buy our oil. We'll deal with the carbon.'

The question of who is responsible for the carbon (or the water loss, or the child labor, or any number of negative consequences of a given activity) permeates international discussions and policy negotiations on trade and on environmental law. But it also crops up at other scales, not just in the international arena, so it becomes important to understand not only who but also what part of a larger process is to be held responsible for mischief, and who or what part gets to bask in favor for a certain benefit.

The complex and often contradictory way of dealing with sustainability factors such as carbon emissions is important to our longer-term deliberations on sustainability assessment. As we work toward goals such as carbon neutrality or even negativity, each intermediate or component element of fuel manufacture will be critically assessed; commercial, economic, social, and environmental needs and interests will become clearer. As that process evolves, the 'you take this, we'll take that' conflicts of attribution will be revealed, and must be resolved. For these and many other reasons, efforts in assessment must be clear, thorough, and as comprehensive as possible within the limits of current knowledge of sustainability factors and their treatment. A sustainability assessment organization must bring as many factors as possible within its standard and treat them rigorously. Ideally, client organizations should be reacting to and participating in decisions about what their certifying organizations want to include, where, and why.

Voluntary, Collaborative Schemes

One criticism of certification systems is that they are usually voluntary. That is only true where there is no legal requirement, of course—a situation that can change. But where compliance is voluntary, whether due to an absence of a regulatory authority imposing a standard or because an industry wishes to meet a higher standard for its own reasons, the stigma associated with any degree of 'voluntariness' should, logically, be defeated. For international air commerce, broad industry groups like the International Air Transport Association (IATA), or organizations whose focus is narrower, such as the Sustainable Aviation Fuel Users Group (SAFUG) can mount campaigns that would make compliance effectively mandatory, not by statute but by agreement. The lower the rate of subscription to a voluntary or voluntarily higher standard, the more aggressively the individual airlines and their sustainability affiliation groups will probably have to publicize the kind and degree of their commitment. Suffice it to say that if the industry's efforts in achieving sustainable fuel are to be considered credible, standards must be developed and applied scrupulously, and enforced rigorously.

Clarity Matters

Given the number of understandings (and misunderstandings) about what sustainability is, we have to accept that any confusion or difference of interpretation is going to work to the detriment of those, like the aviation industry, who are going to depend a great deal upon the degree of public approval and endorsement concerning efforts to quantify and address sustainability matters. The greenwashing radar is now on, and fully alert. It is my strong feeling that the aviation industry should never allow itself to get into a position of having to defend what is being done; that public relations battle just does not seem very winnable. Perhaps the way around that possibility is to engage potential critics in the process of designing aviation's response to its sustainability challenge. Anyone who has a significant profile or media/internet voice can be invited to provide input or, where organizations have the resources, to participate in working out aviation's mechanisms for improving its sustainability.

It goes without saying that an industry with a preoccupation concerning commercial viability is going to have a very different perspective from that of an environmental or social advocacy group when it comes to understanding and assessing sustainability. But unless the industry can engage, listen to, and act on the input of these different voices, it will be very difficult to become the sustainability pathfinder and champion that aviation absolutely needs to be. In fact, one of the most credible efforts in assessing sustainable fuels and developing a certification standard has done just that.

The Roundtable on Sustainable Biomaterials (RSB)

It would be irresponsible to say that the sustainability assessment puzzle has been solved perfectly and completely. It would be equally irresponsible to neglect to describe the positive achievements that have been made. Having tried to examine all the things that present themselves as sustainability assessment resources, I take the balance of this chapter to describe the best that I could find.

Initially the Roundtable on Sustainable Biofuels, the RSB began as a project of Ecole Polytechnique Fédérale de Lausanne (EPFL), a Swiss technical university. In 2013, it became a stand-alone *non-profit* organization, established its secretariat in Geneva, and changed its name to Roundtable on Sustainable Biomaterials; the RSB has also expanded the scope of its certification system to accommodate such materials as bio-plastics, bio-lubricants and packaging. Its credibility and usefulness derive from its commitment to strong, hard, measurable standards, along with its remarkable support by many players from different sectors, including agriculture, environmental and social advocacy, industry (manufacturing, transportation, banking), as well as intergovernmental and non-governmental groups. The project is relatively new and dynamic, and the RSB seems to be the most successful effort at certifying fuels and other biomaterials in a comprehensive and valid way.

The following paragraphs describe the provisions that are outlined at the RSB's website (www.rsb.org). The whole field of sustainability assessment and certification is very dynamic, and the RSB is evolving at the pace necessary to adjust to the situation. For that reason, my best advice is to visit the RSB online for the most up-to-date information. What follows is my understanding at this writing.

Structure

The RSB membership consists of a large number of independent organizations from more than 30 countries. Members are grouped into seven Chambers, which gather together similar voices and interests. The first three Chambers constitute organizations on the commercial and production side:

1. producers of the biomass—farmers and growers;
2. industrial organizations that turn the biomass into biofuel or other biomaterials;
3. associated organizational or corporate commercial interests—secondary processing, finance, transportation, marketing, etc.

The next three Chambers bring together the interests on the social and environmental advocacy side:

4. rights-based non-governmental organizations (NGOs) for land, water, human, and labor rights, together with trade unions;
5. organizations dealing with rural development, food security, smallholder farmer and indigenous peoples' organizations, and community-based civil society organizations;
6. environmental, conservation, and climate change organizations, including policy groups.

The last Chamber is interdisciplinary:

7. intergovernmental organizations and governments, agencies that offer particular technical expertise such as standard-setting or certification, technical or advisory groups, and consultants.

Each Chamber elects a Chair, who leads Chamber meetings; nominates a Director, who represents the Chamber in the Board of Directors (Steering Board); and appoints three delegates who represent the Chamber at the annual Assembly of Delegates, the RSB's highest decision-making body. In this way, the broadest set of views and interests are represented. As a matter of policy, selection of Chairs, Directors and Delegates incorporates appropriate geographical representation from Global North and Global South members (developed and less-developed countries, respectively) as well as ranges of expertise.

The integrity of organizations like the RSB resides in its ability to embody as much legitimate view as possible. On that criterion, it holds itself responsible (just as its clients do) for achieving a standard set by others. For that reason, it maintains membership and adheres to standards set by the International Social and Environmental Accreditation and Labeling Alliance (ISEAL).

While we are describing membership and organization, we should add a few particular observations: The RSB has been extremely successful in involving all sorts of entities to come on board as active participants, and yet saw the withdrawal of certain trade groups involved in renewable biofuels. The point—sensitive, obvious and previously noted—is that organizations that represent and advocate on behalf of particular interests are vulnerable to inherent conflict of those interests with others. Trade groups, professional groups, and other advocacy groups—even whole countries—all have constituencies, but the interests (especially financial) within any constituency can easily be in conflict with the broader effort to understand and fairly address the largest and most significant effects of that group's function. Where the indirect effects of an activity are important in their own right, and possibly in conflict with other specific interests of the entities undertaking the activity, those indirect effects can only be assessed reliably by a disinterested body. This is an extremely important point that the aviation and fuel industries must internalize. The value *to the airlines* (and related industries) of third-party certification has to be recognized.

The RSB embraces its obligation to include in its work as many relevant points of view as possible. But that inclusiveness and variety of views must not turn into mere conflict; it has to become a real, collective estimate of what must pertain. Inevitably, sometimes an organization's constituency will find an accepted consensus principle to be intolerable, but that is in the nature of these efforts. I would say that the RSB stakeholders are far better off together, now that they have found each other in an effective organization, than they are apart (grinding their respective axes.)

Continuing with organization and structure: The RSB acts in partnership with a growing number of independent certification and accreditation bodies, currently based in Germany, Sweden, Australia, and the US, for the management and implementation of sustainability certification tools and mechanisms. Although the RSB now includes non-energy bioproducts in its mandate, even that is pretty narrow. Thinking about fuel, while it was originally organized to certify biofuel, nothing would prevent its systems of assay and authentication of environmental and social value from being applied to any kind of fuel, including alternative fuels that are not biofuels at all.

The RSB is an absolutely unique organization, not in what it sets out to do, but in its singular success in doing it. And I will state plainly that the aviation community should commit to the RSB in a strong and comprehensive way. This commitment should be both broad and deep, working with all commercial air operators to find ways of securing only supplies of RSB-certified fuel to meet alternative fuel targets, and encouraging the RSB to go constantly further in developing the rigor, comprehensiveness, and credibility of its standard, since that credibility will reflect the credibility of the air industry itself.

What the RSB is Able to do

It is a key point that in some dimensions the RSB certifies to a measurable degree, while in others it spells out a 'thou-shalt/thou-shalt-not' kind of limitation. Currently, the native RSB criterion for carbon dioxide is that a fuel blend will be certified if its total carbon content, including the 'bio' component in both its biomass production and processing, results in a fuel that is 50 percent lower than a straight benchmark fossil-derived version of that fuel. So, a jet fuel blend would be certified as sustainable if the total carbon content of the various blend components taken together contain no more than 50 percent of the carbon in the similar 'original' fossil fuel version. But in carbon terms, there are actually *two* levels of restriction and the specifics are interesting:

- a fuel *blend* that is certified as sustainable needs to be carbon-reduced to a specific degree (currently 50 percent), but
- each blend *component* must only be, to some extent, lower in carbon than a fossil-derived equivalent.

Relevantly, this implies that the RSB recognizes not only the specified 50 percent benchmark, but also must know what degree of carbon reduction is achieved by any unblended fuel or fuel blend component. That is important for the application of the certification mechanism to clients who would want to specify certain higher standards.

This type of assessment necessarily reflects typical standards set by statute for some jurisdictions, or typical goals set by user groups. For example, in 2007 the Commission of the European Communities recommended that by 2020, biofuels constitute a minimum of 10 percent of total transport fuel consumption (Commission of the European Communities 2007). The amended EU Renewable Energy Directive (EU RED), 2009/28/EC, reiterated this minimum 10 percent target, and in Article 17 specified that 'biofuels' in terms of that directive must reduce carbon content from the fossil fuel benchmark by 35 percent, rising to at least 50 per cent by 2017 (European Parliament and Council 2009a). So a two-threshold approach to carbon reduction is established: fuels (more likely blends of fuels) must have a lower-than-benchmarked carbon content, and total fuel use must reflect a commitment to those lower-carbon fuels. In that context, it goes without saying that the RSB could respond to the need for any particular level of carbon content certification. It can and does also certify as to the actual content of a fuel, so a fuel blend or blend component will have its carbon content disclosed as part of its documentation.

Note that certification only specifies minimum standards: the air industry could choose to set a more ambitious schedule for achieving a greater degree of sustainability on any particular criterion, including carbon reduction.

The advantage of having a fuel property that is measurable and quantifiable (such as carbon content) is clearly apparent. But in other dimensions of sustainability, this is less feasible. In assessing social justice, how do we say that something is 50 percent just and fair, or 50 percent advantageous for economic advance, or 50 percent of some other abstract quality that we consider to be part of sustainability? In those cases, it is more useful to ask a simple yes or no question: Does this plan comprise provisions that are going to advance economic or social prospects for those affected? Does this plan insist on fairness in treatment of those same people?

As well as considerations that are not readily quantifiable, there are others that may lend themselves to measurement but whose threshold of acceptable harm is zero. These things must also be assessed as either acceptable or not without making any grey-scale determination.

The RSB itself cannot impose a legal sanction, but it can operate to provide a mechanism for certification in favor of a state entity (or any other entity for that matter). If there is a state regulation for reduction of carbon content of fuel, the RSB can license certifiers that will ensure that the fuel meets the state standard and is sustainable in other respects. It is noteworthy that in June 2011, specific standards developed by the RSB were accepted by the EU RED (European Commission 2011).

The RSB Global Standard for Biofuels

So, what *is* the standard? The RSB has published a number of documents, most available in various languages and many also prepared in versions specifically tailored for application to the EU RED or to the German market.[2] *RSB Principles & Criteria for Sustainable Biofuel Production* [RSB reference code: RSB-STD-01-001 (Version 2.0)] (Roundtable on Sustainable Biofuels 2010) is the 'global' or generic RSB standard for production and processing of biomass into usable biomaterials. Although published in 2010, the introductory paragraphs mention criteria and procedures expected to take effect in 2011 and 2012, and explicitly state that the standard must evolve, integrating new information and technologies in order to stay relevant in the rapidly changing biofuel sector.

Principles and Criteria

In the RSB standard(s), principles that outline a specific requirement are matched with appropriate criteria that clarify how the principle is fulfilled. The twelve

---

2   See http://rsb.org/sustainability/rsb-sustainability-standards/ for the current list of documents available for download.

principles and main criteria of the global standard are described below, but for convenience only. Since this is a rapidly evolving initiative, the website must be consulted for new and updated information. Not only that, but the actual detailed principles and criteria as published are far richer and more impressive than any summary can convey.

One other *very important point*: If we know that the principles and criteria will develop and evolve, we can fall into the habit of extrapolating from current provisions only, imagining a greater degree on a certain criterion. Remember that the evolution will also see the introduction of *new* criteria. For example, provision for land use change (ILUC) and residues are in the process of being introduced at this writing.

I present the standards and criteria (mostly paraphrased) with some brief observations incorporated here, and then some more detailed comment further on.

1. Legality

The first principle is simply that biomaterials operations should comply with local and international law as applicable.

2. Planning, Monitoring, and Continuous Improvement

Let us note here that the overarching 'mother of all criteria' is that fuel (or other) initiatives must have a social license in order to proceed. This consideration is different from the comprehensive requirements for environmental benignity and economic viability or advantage, in that the assessment of the appropriate application of each principle and criterion demands permission and approval, which absolutely necessitates prior, informed consent.

Consider how essential an element this is in terms of making any other standards and criteria relevant. Not every single consideration that is important to individual stakeholders will be addressed by the standard. That will always mean that stakeholders need to make their own determination as to whether a principle or criterion is sufficiently and appropriately applied in a way that captures concerns that are not more clearly or specifically spelled out. So the principle of *free, prior and informed consent* (FPIC) is a crucial element in the application of *all* criteria. To this end, people need time to inform themselves, full information to work with (including active consultations and counseling—people may not know what they need to ask about), and a complete absence of coercion in coming to their decisions. The people who are affected most directly by such changes as occur within their setting of life and landscape must be satisfied with the way things are going to be.

So to return to the second principle: it relates to constant examination and review of projects to make sure that they are compliant and improving. Criteria 2b and (later) 12b explicitly mention 'free, prior, informed consent' as the basis for consultation processes and negotiated agreements, but the concept is fundamental to the whole document.

Looking at the principle itself, the terms *planning, monitoring*, and *continuous improvement* clearly point out the need for impact assessment, and risk and economic viability analysis, which should all be consensus-driven and gender-sensitive to arrive at worthy negotiated results. The criteria specify what sort of environmental and social impact assessments might need to be done. The requirement of free, prior, informed consent implies the active involvement of all stakeholders in determining the benefits and risks in all dimensions, both initially and ongoing.

3. Greenhouse Gas Emissions

This principle enjoys a particularly high profile, given the attention to the climate crisis, and is of particular interest to us. The criterion here is that carbon reduction be measured in a way that essentially abandons the concept of system boundary, by embracing a carbon accounting that extends from 'well to wheel', including land use issues such as the change in below- and above-ground carbon stocks; in this comprehensive manner of carbon auditing, the use of co-products and waste is incentivized. Criterion 3c spells out that fuel blends must have a 50 percent lower carbon content (and blend component provisions as discussed) than the fossil fuel baseline, but must also meet the standard set by any jurisdiction entitled to set one. All of this would have to be undertaken on the basis of appropriate life-cycle assessment that captures the relevant flows of carbon from the air, through the processes of production, transportation, processing, and manufacture, right to the (in our case) airplane. So these calculations need to account for the use of fossil fuels in machinery, or vehicles, or the production of fertilizers, for example. The carbon content baseline is the carbon in an equivalent fossil fuel.

4. Human and Labor Rights

This principle and its associated criteria are immensely powerful when one considers the application of the FPIC provision.

Human rights, labor rights, decent work, and worker well-being: we can imagine that in many circumstances, these provisions will be foreign to the point of being strange to those whose endorsement is sought. Here is an opportunity for making a profound difference. Regardless of where located, the workers in an RSB-certified operation must be able to associate, organize themselves, and bargain their arrangements as they see their interests. Forced or slave labor is prohibited, of course. If children are involved, it can only be on family farms and only if they are also allowed to be educated and to work in a physically healthy way and environment. No discrimination of any kind—equality with respect to gender, opportunity, employment, wages, working conditions, and social benefits. Compliance with laws and international conventions is required, as is adherence to collective agreements and to international standards for occupational health and safety. All of this applies even to contracted third-party labor.

5. *Rural and Social Development*

In regions of poverty, RSB-certified operations must assist with the social and economic development of local, rural, and indigenous people and communities. The socio-economic status of all local stakeholders must be improved. The participation of women, youth, indigenous societies, and the vulnerable must be catered for.

6. *Local Food Security*

RSB-certified operations have to ensure the human right to adequate food. They must assess existing food security risks and mitigate any negative effects from their own operations. They must also ensure that food security is improved in food-insecure regions, and enhance the food security of directly affected stakeholders.

7. *Conservation*

Operations are only to be conducted if they avoid negative impacts on biodiversity, ecosystems, and conservation, and maintain or even improve related values of local, regional, and global importance, along with ecosystem functions and services. Habitat is preserved; invasive species are prevented.

8. *Soil*

Practices must reverse any degradation and maintain soil health, including physical, chemical, and biological conditions.

9. *Water*

The principle states that biofuel operations shall maintain or enhance the quality and quantity of surface and ground water resources, and respect any prior formal or customary water rights. The criteria specifically mention the water rights of local and indigenous peoples, and the establishment of a water management plan that uses water efficiently, maintains or improves water quality, and does not contribute to depletion of surface or ground water.

10. *Air*

The principle is clear and concise: Air pollution is to be minimized.

11. Use of Technology, Inputs, and Management of Waste

Technologies should maximize efficiency and minimize harm. Some criteria here are very interesting. The first makes clear that the law can impose limitations on our ability to ensure FPIC: Criterion 11.a Information on the use of technologies in biofuel operations shall be fully available, unless limited by national law or international agreements on intellectual property.

But many of the remaining criteria specify that (even taking into account that intellectual property law may prevent detailed knowledge) use of genetically modified organisms (GMOs) shall be undertaken on the basis of minimization of long-term risk to environment, people, and their society. The last two criteria are more pedestrian if equally important, relating to handling of chemicals, products, by-products, and waste.

12. Land Rights

'Biofuel operations shall respect land rights and land use rights.' This principle goes right to the heart of one of the most important elements of social justice. I'll quote the criteria as well, in full:

> Criterion 12.a Existing land rights and land use rights, both formal and informal, shall be assessed, documented, and established. The right to use land for biofuel operations shall be established only when these rights are determined.

> Criterion 12.b Free, Prior, and Informed Consent shall form the basis for all negotiated agreements for any compensation, acquisition, or voluntary relinquishment of rights by land users or owners for biofuel operations (Roundtable on Sustainable Biofuels 2010).

Additional Commentary on the RSB Principles and Criteria

In aggregate, the RSB principles and criteria constitute a substantial milestone in bringing the theoretical idea of sustainability to life. This starts to become particularly apparent when we examine Principle 4, which relates to labor and human rights. Not only does it support the rights of workers as enshrined in the International Labour Organization's *Conventions, Protocols,* and *Recommendations*[3] but in some ways surpasses them. It is interesting, and illustrative of the RSB's approach, that the standard will not tolerate abrogation of RSB principles on the basis of a state's laws: where a state restricts, for example, freedom of association or right to collective bargaining, the RSB criteria require that an operator figure out a way to afford its workers these rights in a manner that is not in conflict with state laws and

---

3 For a current list of ILO publications, see http://www.ilo.org/global/standards/information-resources-and-publications/lang--en/index.htm

then say, effectively, 'Over to you.' So regardless of how little protection the state allows, fuels will be certified to the higher standards. This means that companies and governments must study their interests and either make their decisions on the basis of RSB principles, or abandon the opportunities that certification affords.

The eleventh principle, Use of Technology, starts to tie the other principles and criteria together more thoroughly. The principle that technology must help rather than hurt efforts toward sustainability takes us back to our earlier discussion about the need to make technology perform in its pursuit, but to be wary of the potential weaknesses of simplistic approaches that allow the application of technology to create more, new, and bigger problems. Criterion 11.a again supports the need for free, prior and informed consent by requiring that the information related to any technology and its use be made entirely available.

Principle 12 comprehends the simple and profound truth that the land gives definition and confers reality to people and their societies. The effect of emissions of greenhouse gases involves the entire planet, but for all other things, the land defines the considerations. A society's land may represent everything about themselves that they want to preserve and protect, and patterns of land use are fundamental to identity, worth, and meaning. Consequently, land rights and use entreat appreciation that goes beyond legalities. Consider, then, how challenging it might be to engage people in learning about whether or not they might want to change all of that in order for a project to go forward, when they have no experience or abstract knowledge of the *possibility* of changing it. Free, prior and informed consent has particular resonance in such situations.

Comments From Barbara Bramble

The foregoing as it pertains directly to RSB is based upon material that is freely available on the RSB website. But I spoke at length with Barbara Bramble, who directs the National Wildlife Federation's program on International Climate and Energy, and serves as Chair of the Board of Directors of the RSB. Her view is that the evolving effort in developing the RSB principles and criteria constitutes an effective 'definition' of sustainable materials production. And I have argued that it seems, absolutely, to be the best thing that we have in that regard.

That does not mean that all will embrace the standard. Because it is rigorous, those who cannot meet it will demur and therefore trade associations, whose memberships comprise a substantial number of such producers, may not wish to be associated with the RSB directly. They must serve the needs of their member producers. That is why trade associations are not relied upon in that way. It is *users* who drive the advance of such efforts. Airlines will want to be able to secure supplies of fuel that can claim a credible sustainability certification. Individual fuel suppliers will seek certification to meet that need.

Bramble pointed out that the work is never going to be over, and even big things remain to be addressed. She confirmed that the RSB's new policy on ILUC still needs to be embodied in specific standards text; new categories of by-products and co-products will be identified; and GHG reduction thresholds have not been calculated for bio-plastics and other biomaterials. They are not being neglected.

Speaking specifically of the question of co-products, she pointed out that it was not acceptable to give such materials a 'free pass'. It is generally not possible to certify the whole production chain for such materials either, especially where the *principal* product may have no need of such certification. It may also be difficult to identify the upstream producers, or to allocate specific negative effects through such a supply chain. If those three things just mentioned cannot be done, what *can* be done? Out of an unrelated and unassessed product chain, pops a useful by-product/co-product. Surely it should not go to waste. And even if it has some other commercial value, if the highest and best use is in production of a biomaterial, should it not be put to that purpose? Bramble raised the example of beef tallow that may be produced through agricultural and food processing chains that are not involved in certification. Such tallow might have an otherwise low commercial value but be a *very* valuable sustainable fuel component. Therefore, the RSB chambers were able to agree on a simplified process for certifying both residues and by-products. 'Residues' are things that we would formerly have called 'waste'.

Of course the RSB mandate and 'sustainable biomaterials' is a much bigger matter than aviation fuels, but Bramble, like many others, is attracted to this particular realm. She is drawn to the nature of the quest: Within the planet's demand for portable liquid fuels, commercial aviation constitutes a manageable small part of that total. And it is work that seems doable. It is a 'high value compartment' with motivated participants. She sees this as helpful where, in the context of the total global requirement, there is a need to start modestly and carefully. She notes, as others have, that commercial aviation fuel is a great place to start.

I asked Bramble what she thought of application of the standard to other sorts of materials that are pursuing sustainability with no real 'bio' element to the materials involved. She had no qualms that RSB would be ready to operate outside of any narrow delineation implied by its name. And she pointed out that sustainability certification would have to become effectively universal and so a proliferation of certification organizations was probably not useful when the RSB and a small number of similar dedicated groups could otherwise fill the bill.

The Air Industry's Situation: Applying a Standard

There are clearly great challenges ahead. Much will depend—and it may not be going too far to say that *everything* will depend—upon the rigor that the aviation industry brings to certification processes. Every important bit of objectivity, credibility, authority, reliability, and integrity will be essential in determining whether we have processes that can actually tell us something reasonable about how we get these new fuels.

Like the teenager, all we need to do is pick up that first sock. Can every airline in the world and every flight—domestic or international—start up on a given date and be perfectly 'sustainable'? No. It is appropriate to emphasize that choosing a method and set of gauges, along with one or two certification bodies to apply them, does not mean that all sources of fuel will meet the same degree of sustainability at the same point. Again:

- jet fuels must be fungible (able to substitute for each other),
- any airline can buy any fuel, anywhere, while actually burning some other fuel, anywhere, in substitution,
- national standards of carbon content for domestic air travel can vary from country to country,
- standards for international flying may be different from all national domestic standards, and
- the ability to lower the carbon content of jet fuel may vary from place to place.

All of this permits and encourages initiatives to integrate the demand for fuel and the demand for jet fuel carbon reduction into a useful global market that accommodates these values, *as long as the carbon reduction is measured in a consistent way*. The advantages of exploiting jet fuel's required fungibility, and allowing each entity and jurisdiction to meet its own particular goals and standards are hugely important to everyone and to the climate change problem.

As exemplified in the Kyoto Protocol, countries (including their domestic air operations) can accept respective amounts of responsibility for reducing carbon dioxide emissions; international air operations can commit to other obligations. So an organization for sustainability assessment and certification does not need to impose any specific carbon reduction value worldwide, it can perhaps simply certify one fuel as, say, 'C-50' and another as 'C-55' and let the fuels be bought and burned as appropriate. What is important is that the certification methods and measures be perfectly uniform, and that a possibility exists for a carrier to acquire a contract for fuel of the desired 'C' value. So we need standardization of ways of assessing sustainability and the specific carbon content of blends, and we need markets to move carbon reduction to where it has value. This means that airlines need to submit to a common standard-setting methodology, and the easiest way to do that is to allow a supranational body (or two) to handle at least this one thing. Desire for national control of important matters is understandable, but this impulse must be overcome if there is to be any credibility attached to the standardization effort. If there are to be separate national certification bodies, they must still each certify in *exactly* the same way.

Evolution

Thinking about these standards and criteria makes one wonder where such efforts will take us. The RSB standard, with its principles and criteria, certainly constitutes a remarkable achievement. But it is not akin to some edifice—standing there finished and waiting to be accepted and admired. Just like 'Sustainability' itself, the RSB is a project. It seems clear that it is not 'finished'. That is not a weakness; it is in the nature of our collective task as we have considered it.

And it is a project that may grow and spread. Appreciating what the RSB has accomplished and what it hopes to accomplish, can this way of assessing be applied more broadly? Can it be reproduced or even improved when others bring similar care and thoroughness in other industries and activities? That seems likely. We have described how systems of life-cycle analysis need to probe into materials and process pathways to understand if the inputs—the materials, energy, and processes that provide what is needed to make sustainable fuel—are, themselves, sustainable. Just as the air industry has many clients, so it and the companies that will manufacture its sustainable fuel are clients to a myriad of goods and services providers. As well, consider the matter in the context of remaining *within* the air industry: if we make this kind of effort to ensure that flight energy (responsible for 30–40 percent of air transport's operating costs) becomes more sustainable, the departments that answer for the remaining 60–70 percent will, presumably, have to answer the same suite of questions.

With this chapter, we come to the end of understanding the nature of the task at hand. Now, we will start to think a little bit about the keystone in the arch: doing what is necessary to enable and encourage real production of sustainable flight energy.

Chapter 11
Policy Development

There are people making fuel that they consider sustainable. Some of it is getting certified. There are airlines buying a bit of it. The difficulty is that, by any reasonable standard, the rather amazing new technologies that actually give us fuel that advances the cause of flight energy sustainability are not making inroads at anything like a sufficient pace. There are many ways to push things that perhaps need a shove—advocacy and consumer action can be powerful forces. But these can remain diffuse when it comes to an activity that is conducted by hundreds of companies in hundreds of international and domestic markets. Policy is the tool that can bring some effect to public will, so we will examine policy at both the international and domestic levels. Whether a person works in industries related to air travel or is concerned about their own individual travel carbon profile or that of their organization, if change is desired it is to the policy mill that we must look. While real policy development is an immensely complicated and fraught enterprise, the rule-makers need to know not only what their policy will attempt to accomplish in theory, but also how others understand the policy challenges and react to challenges and policy ideas, both. Bluntly, and in preface to the rest of the book, anyone who wants to see change should advocate and lobby heavily for policy action even as they modify their own activities and campaign publicly to sway others.

We described the policy history of international aviation's commercial aspects and also international climate negotiations in Chapter 5. That was part of the 'background' or 'setting up' portion of our inquiry. Now that we know more, it is time to answer the most relevant questions concerning policy development. It is a complicated subject, because the naked factors interact in myriad ways: a price, penalty, obligation, subsidy, or advantage created *here* affects everything *there*. But even that reality does not consider the further level of complexity that results when politics are added. Everyone has their own vision. Our discussion does move from an attempt to aid comprehension toward new ideas about policy, but it is not really intended to offer concrete policy proposals, per se, just a reflection on ideas that stand in contrast to existing shapes of policy that are known to be somewhat (or entirely) problematic.

Policy Pitfall: Stepping in the Wrong Direction

We return to the caution that we outlined earlier: Bad policy begets worse policy as the errors start to become apparent and their effects compound. I am choosing a recent example that has nothing to do with aviation so that it can be considered as objectively as possible. It concerns electrical solar power in one country.

Policies related to the promotion of residential rooftop solar photovoltaic electric power production in Spain seem not to have anticipated very well the way in which markets and major power utilities would be affected by their introduction. The intention was evidently to encourage the use of renewable energy, but there were consequences: utility companies suffered. Not only was there reduced demand for the conventional (grid) electricity that they sold, but households were allowed to sell their *excess* solar-generated electricity to the grid at premium rates. Conventional utilities struggled with reduced revenue, and the cost of financing the public subsidies for renewable energy was crippling (Roca 2013, Pentland 2013). Disregarding the details for now, it is instructive to note that its earlier policy decisions have resulted in Spain's cash-strapped government being forced to reverse itself, now imposing tariffs and penalties upon residential producers of solar power, thereby hurting the citizens who took advantage of a government program in good faith, and who had no way of imagining that there would be financial exposure for themselves and their families. As long as a household maintains a working solar array, any electricity that they *buy* from the utility is billed at a higher rate. If they disconnect a functioning array from the grid and use solar exclusively, they are fined. One now expects that some Spanish homeowners will actually dismantle their very useful solar arrays as a result. The Czech government finds itself in a similar quandary (Vorrath 2013). Could either government have anticipated the current situation? We cannot say, and it is easy to criticize—but I would bet that they wish they had. It is patently ludicrous that better, in-place power generation capacity, hugely difficult to win in the first place, would be forced out—not only off the grid but out completely. Similar policies have been used elsewhere with good results, what was missed *here*?

Unless everyone gives some thought to how policy is being developed, or *could be* developed, it will not develop properly. In our aviation context, we have to talk about policy development in terms that everyone can understand, and how we might best proceed toward the specific goal of producing jet fuel in sustainable ways.

The immediate, real-world problems that confront politicians and policy-makers are often big. Sometimes they are so important, and their practical or political immediacy so prominent, that the policy response tends to become shortsighted and loses comprehension of an essential element: policy is supposed to guide strategic approach, not serve immediate wants, political or otherwise. When policy development responds to a situation rather than to an ethos or goal, it can create policy traps. Sustainability will be challenge enough without all of the mis-starts that would likely imperil the whole venture.

What Needs to Get Done

Just as we wanted to know the size of the challenge when we were talking about our technical options for sustainable fuel, we should think again about scale as we consider the policy that will help. As we already know, the challenge is daunting. And this book is not an attempt to express what is likely to happen in the normal course of events, applying the usual old ways and means of trying to make things a little better in the world, it is written in the context of what *has* to happen. That is considerably more important and impressive than what a reasonable person might expect in more reasonable circumstances. So I do not claim that aviation's sustainable flight energy ambitions will be realized, nor do I predict that policy will evolve in a way that provides real, material benefit toward that realization. But nothing is gained by understating what needs to be accomplished in order to make it seem more realistic or plausible. If the goals that are described in these pages seem far-fetched, I ask you to remember what is at stake.

On Giving Up

Many commentators or researchers already tell us what is 'likely' to happen in aviation's pursuit of sustainable flight, and such analyses are valuable as alerts. In their 2013 report *Mitigating Future Aviation CO_2 Emissions: 'Timing is Everything'*, Lee, Lim, and Owen foresee that the industry will achieve rather modest reductions in aviation emissions and global warming effect over the next few decades to 2050—at least they seem modest in terms of approaching zero. The report also states that market-based measures (MBMs) such as emissions trading systems offer the most efficient CO_2 reductions, and should be the backbone of our efforts (Lee, Lim and Owen 2013b).

However, this 2013 analysis assumes a 'likely' penetration of perhaps 10 percent by biofuels over the period to 2050, using data from a 2009 report that notes, 'Concerns about land availability and sustainability mean that it is not prudent to assume that biofuels in 2050 could account for more than 10 per cent of global aviation fuel' (Committee on Climate Change 2009).[1]

Such surveys of past and current progress in implementing alternative technology would normally lead us to limited expectations, but they also ignore the fact that we have to achieve much, much more, no matter how extreme that sounds. If we agree with these researchers that cap and trade is likely to produce the bulk of any progress that we see, we risk making that response the matter's destiny by not examining other possibilities. MBMs will certainly have a role to play, but sustainable, alternative sources of fuel—no matter how challenging—are the only options that offer any potential of actually accomplishing enough in

---

1   The 2009 study also examined the potential use of hydrogen energy but concluded that there were technical and logistical reasons for not including it as sustainable alternative fuel in the immediate future.

solving the real problem. The Lee report is an immensely valuable tool. But it does not provide the essential basis for policy: policy's job is not to extrapolate from the present in the direction of something that has a clear shot at solving *part* of the problem, it is to set a new course into the future *as required by that future*—to solve *all* of the problem.

In Chapter 6 of Sir Arthur Conan Doyle's literary classic *The Sign of Four*, Sherlock Holmes explains one bit of irrefutable detective logic to Watson: 'When you have eliminated the impossible, whatever remains, however improbable, must be the truth' (Doyle 1890). We can make a similar sort of case here: if we identify all of the things that we might continue to do to reduce flight emissions of GHGs, and all of those things do not add up to a sufficient reduction, then whatever single thing that *would* allow us to succeed must be pursued through to the desired result.

Airplane and airspace efficiencies may do their part. But, absent sustainable fuel, MBMs cannot cover the rest of the distance. We cannot move an emissions cap toward zero unless there is a physical way of achieving that zero, notwithstanding all of our options for buying and trading carbon credits in the interim. And neither the industry itself nor any country wants an inexorable reduction toward zero flying as the alternative. That would be the implication of such an approach. Unless we have sustainable fuel, we would ultimately force every airline to simply buy carbon offsets. But if every other part of the global economy is striving toward the same end, from whom will airlines buy credits? And how much will they cost?

Ultimately—and I stress that word—we are stuck. And however far in the future 'ultimately' might be, the real goal has to be clear in our sights now so that we do not send ourselves into *culs de sac*. No matter how difficult it seems, and no matter how inappropriately it has been pursued up to now, lowering the actual carbon content of the fuel, while simultaneously ensuring that its production satisfies the other aspects of sustainability, is the only solution that answers.

The Quick Versus the Slow Start

Lee, Lim, and Owen (2013b) make a very good case on another previous point worth emphasizing: As we have discussed, the carbon that goes into the atmosphere acts as a greenhouse barrier the whole time that it is there, and it has a long residency. When we pursue reductions in *rate* of emissions in order to arrest the growth in the accumulated *amount*, the speed with which we reduce the rate determines how much carbon we live with in the end. If we were to achieve a total, global emissions rate of zero *today* for every single state, industry, and individual, we would effectively arrest the atmospheric level of CO_2 at its current level, about 400 parts per million (PPM). Accomplishing emissions rate reduction goals more slowly leaves us continuing to accumulate atmospheric carbon stocks in the interim; by the time that we arrest the rate of emissions, accumulation might be 500 or 600 PPM or higher—it all depends upon how slow we actually are. Even though we may create an entirely new energy regime for aviation in the future (such as hypothetical sustainably sourced hydrogen), we absolutely need to bring emissions down now, anyway. In that context, since we know that it is possible to source sustainable conventional jet fuel in other ways, getting started is essential.

Figure 11.1 provides notional examples of how achieving emissions rates of zero at different times between 2020 and 2050 leaves us with different levels of accumulated emissions and the climatic effects that attend higher levels.

Figure 11.2 illustrates more generally the relationships between emissions rates and corresponding accumulated levels of CO_2.

Noting again that these are visual approximations, the horizontal lines representing unchanging levels of accumulated atmospheric CO_2 upon achievement of zero rates should actually trend down slightly to represent the very slow natural processes that remove CO_2 from the atmosphere—unless humanity raises levels to the point where other effects take over, and the levels continue to rise all by themselves. These relationships are well understood by most people—but not by everyone. Their recollection is critical to assessing various policy paths.

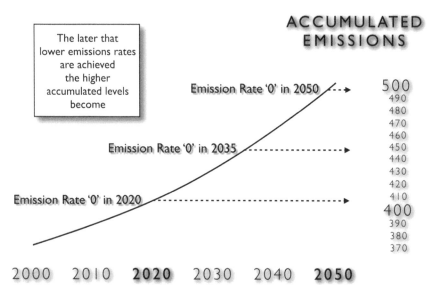

Note: The horizontal dotted lines represent the hypothetical achievement of a global rate of emissions of zero

Figure 11.1 Illustration of the effect of achieving zero global emissions rates at different times

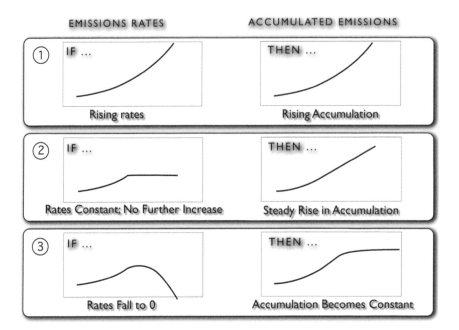

Figure 11.2   Relationship between greenhouse gas emission rates and accumulated amounts in the atmosphere

Remembering 'Non-carbon' Effects

Since we are discussing the matter of long-term evolution to ideal solutions, we have to consider aviation's total flight effect. We must not forget that factors other than greenhouse gases are important, and policy development efforts must not ignore their existence. We talked earlier about non-carbon aviation emissions factors such as contrails and aviation-induced cirrus (AIC). We need policies that will address those as well. AIC is not a 'commodity' that is easy to monetize. MBMs cannot easily be applied to this part of the challenge.

A 2011 conference presentation by Ulrich Schumann describes extensive studies of aircraft efflux condensation phenomena, suggests strategies that could reduce their climate impact, and also examines the associated tradeoffs of increased fuel consumption (CO_2 production) and expense (Schumann, Graf and Mannstein 2011). If, indeed, our principal strategies for dealing with such phenomena are in the nature of planning around the routes and altitudes where they occur, fuel burn will be increased. Employing low- or no-carbon fuel may become even more essential. If we must burn more fuel in the effort to address the AIC/contrail problem, that is yet another reason for it to be low in carbon.

The 50-50-50 Challenge: Doing the Math

If we accept that creating new, sustainable sources of flight energy is essential, we should get back to understanding the scale of carbon reduction that we will attempt. Because we are about to plunge into a discussion of what different jurisdictions and bodies should do on the policy front, we need to resolve matters that some of the principal actors wish to keep in dispute. If one voice says that a certain policy measure is efficient and sufficient, while another says that it falls far short (and we see that kind of argument all of the time) the *size* of the task is critically important.

In the broadest terms, we can say that the magnitude of the required reductions in global GHG emissions from all countries and all sectors is great. The science says that we should bring the rate of emissions to very low values, as soon as possible. The distance to be covered is shockingly large. We have to be frank about that.

We are not, in fact, working on eliminating global warming, we are now merely trying to keep the amount of inevitable warming from exceeding some (somewhat arbitrary) level of danger—perhaps 2 degrees Celsius. In that context, in 2010 UN Secretary-General Ban Ki-moon issued a '50-50-50 challenge': By 2050, the Earth's population will have increased by 50 percent (to approximately nine billion people, as compared with six billion in 2000), and we need to reduce global GHG emissions by 50 percent (using the same timeframe) if climate change is to be kept in check (Ban 2010).

This means reducing global average *per capita* emissions by two-thirds. That is because if emissions increase at the same rate as population, their levels would rise to 150 percent by 2050, whereas they need to be at 50 percent. Fifty percent instead of 150 percent—one-third (of the 2000 rate).

But the world is not a homogeneous mass of equally wealthy people. Developed-world (or 'reasonably wealthy person anywhere') *per capita* emissions are far above the current average, and poorer people cannot appreciably reduce their emissions, which are already quite low. We could therefore characterize the challenge as: The rate of emissions of less-wealthy people should not be allowed to increase much, and the rate of emissions attributable to everyone else must drop *a lot*. That means *more* than two-thirds. Or we could be even more honest: If emissions in the developing world start to grow—the normal outcome when everyone works toward improving the lives of those less wealthy—the emissions of those who are already enjoying a good standard of living, and the unit emissions of the activities that constitute that standard of living, must drop hugely.

As we have noted previously, regardless of where they live, people who use commercial flight services are generally among the relatively affluent, and their passenger flights constitute an enormous part of their personal emissions profile. Even where aircraft are used to transport certain materials produced by people who are not so well off, the cargo itself represents wealth, and is shipped to serve the wants of people who can afford to pay the embodied cost of transporting the goods they value. So aviation is right at the center of the need to achieve the huge reductions in emissions required of the relatively rich, both current and future.

The industry's own commitment is worthy. In fact the air industry's international aviation goals of carbon-neutral growth (CNG) by 2020 and a 50 percent reduction in absolute emissions amounts by 2050, as resolved at the 69th Annual General Meeting of the International Air Transport Association (IATA) in June 2013, satisfy Ban's formulation (International Air Transport Association 2013). But it is worth looking a little more carefully at Ban's emissions reduction aims in the context of development and disparity in wealth.

Taking into account the general support of development goals for the world's poorer peoples, we should consider that a 50 percent increase in the world's population is likely to incur much more than a 50 percent rise in economic activity. Absent new ways of doing things, the emissions rate is compounded by the rise in degree of total development.

Recalling the formulation for total emissions:

Population × GDP per capita × unit GDP energy × unit energy emissions

If population rises by 50 percent, and GDP per capita rises as well (or even remains fairly static), the energy intensity of economic activity and the emissions intensity of that energy need to fall, precipitously.

If we are to reduce per capita emissions to one-third of the recent rate by 2050, we must reduce to an even smaller fraction both the energy intensity of economic activity and the emissions intensity of the energy exploited in the cause of that economic activity. In other words, pursuing economic interests must involve ever-decreasing amounts of energy that has ever-decreasing carbon-emitting properties.

A glance at the 'World' data in the CIA's *World Factbook* entry for 'Economy: GDP - Real Growth Rate' shows growth levels gradually declining over the past few years, from (est.) 3.8 percent in 2011 to (est.) 2.8 percent in 2013 (Central Intelligence Agency (US) [continuously updated]). Assuming a not-unreasonable notional trend of 3 percent would yield an approximate threefold increase in global level of economic development by 2050. All things being equal, in order to just maintain emissions at current global levels, we would need to compensate that threefold increase in economic activity with a commensurate drop to one-third of the current unit GDP *rate* of emissions. To meet Ban's challenge of an additional 50 percent reduction, the rate would instead have to drop to about one-sixth of current levels. A more pronounced shift in emphasis or success in raising the development standards of the poorest would probably result in a greater than historic average annual increase in global GDP, which would demand a correspondingly even smaller fraction for rate of emissions.

Thinking about the whole matter of economic growth, disparate or otherwise, we know that we are dealing with very long-term outlooks using economic modeling that is almost certain to err in various ways,[2] so we should not get too fixated on a particular set of numbers, nor should we expect or presume precision here. Instead, just look at the general shape of the likely evolution: The world population rises, and economic initiatives for poorer countries continue; a greater proportion of the world's increased number of people participate in activities that are associated with greater degrees of development. But there is general consensus that *per capita* emissions need to drop to very low levels. Since developing countries' emissions are *relatively* low, developed societies have to take on the responsibility of dropping their emissions in a way that is even lower on a unit basis. The typical activities of a developed society now have to take place in a way that results in just a tiny fraction of the emissions that are typically associated with them now.

Aviation is a very energy-intensive activity that corresponds closely with an individual's participation in developed-world activities—whether that person is a developed-world citizen or not. So it follows that we must certainly achieve substantial reductions in the amount of energy used in each bit of flying that we do. Also, the emissions intensity of the energy used in that air travel must decline to extremely low values. We can calculate some theoretically justifiable very small fraction of our current rate of flight emissions and pursue that as our nominal goal—which would be an important exercise—but for relevance here, it is more reasonable and useful, in my view, to think in terms of reducing flight energy emissions to as close to nil as one can imagine.

Efficiency

We have mentioned several times the part that will be played by efficiency increase. We do know that the energy intensity of flight will drop substantially as the system gets progressively more efficient; airplane design is more functional every day, operational procedures become more streamlined, and we can hope that airspace management will improve too. However, the increase in flying activity outstrips those gains. If we suppose that air industry growth rates continue on the order of 5 percent per annum (as we discussed in Chapter 1) then an approximate sixfold increase in traffic will result by 2050. Improvements in system efficiency (as we outlined in Chapter 2) are likely to reduce energy use substantially (Green 2009, Rutherford and Zeinali 2009), but it is still conceivable to foresee perhaps a fourfold increase in emissions, if the same fuel is used. To cut that (fourfold) 400 percent amount of absolute emissions down to 50 percent, as the industry has pledged to do, the emissions intensity of fuel must drop to one-eighth of current levels—again, a very small fraction indeed.

---

2 A useful reference here is *Insights Not Numbers: The Appropriate Use of Economic Models* (Peace and Weyant 2008).

Can Aviation Emissions Be Excused (at Least for Now)?

Are there any magic escape hatches for aviation in this regard? In larger discussions about policy related to emissions reductions, many commentators muse about priorities and wonder whether, since aviation constitutes such a relatively small source of emissions, policy should favor getting low-carbon fuel into other modes of transportation as a matter of more benefit, and dedicating fossil energy for commercial air travel.

Setting aside, for a moment, all of the practical, political, and technological challenges that would be associated with such an approach, let us consider a more useful perspective. Even if the proposal were otherwise feasible, it seems to me that well-reasoned policy would first reflect two factors:

1. many other transport modes already have a variety of other energy options, and
2. aviation's growth stands to be much more rapid than that of other activities on a global basis.

Aviation will not remain the relatively small emitter that it is now; if policy does not support aviation's rapid migration to a low-carbon profile, it threatens to become a bigger problem fairly rapidly.

An argument can be made that any reasonable hierarchy for investment in carbon-reduction technologies should rank industries on the basis of the importance of the economic activity and development they enable, the energy intensity of the service they provide, and the emissions intensity of the energy they use. Aviation would achieve a very high standing on such a matrix.

Market-based Measures as Bridge

Though I have been at some pains to point out that MBMs cannot solve the problem, they should not be dismissed. There is little support for the idea that efficiency and new fuel sources will keep international aviation emissions from progressing right through their levels in 2020. So carbon-neutral growth from that year onward would be impossible if international commercial flight had no means of closing the gap through temporary access to carbon offsets or credits. This 'bridge' function of MBMs is important and we will return to that matter to see if and how we may succeed there. But for now, that is the outline of the policy challenge as it relates to fuel.

Some Goals and Characteristics of Good Policy

That brings us to the question of what good policy looks like. We now know that sustainability of human action and enterprise is an immensely complex and interwoven fabric. Policy development needs to be as subtle and deft as the subject requires. In the rest of this chapter we will examine some mooted policy characteristics and see why they might help. They respond to some of the obstacles that we saw arise in Chapter 5, where we discussed the tension between the provisions of commercial air policy and global climate policy evolution. In addition, they are intended to support the real provisions of sustainability as it is now understood and as it needs to be measured and assessed. These possible characteristics are brought forward now so that they can be borne in mind as the book progresses to consider how policy is made and what is possible, together with the things that arise when we hear the views of air and fuel industry actors. A brief list follows immediately, by way of introduction, but an understanding of why they were chosen will flow *from* the discussion of them.

First, we can propose some of the considerations that are most tightly tied to making policy that will cause an action, or activity to *be* sustainable. They could relate directly to fuel, for example. Here are three:

1. Employs long-term perspective and identifies ultimate aims
2. Demands absolute uniformity of measure
3. Pursues universal availability at uniform price.

The following are an additional nine characteristics that relate to broader policy, flowing from approaches that policy development takes:

4. Views sustainability as an integrated concept
5. Provides the carrot
6. Complements the carrot with the stick
7. Internalizes effects
8. Establishes relevance in the economy
9. Discloses costs
10. Directs revenues toward solutions
11. Provides some immediate benefits with minimization of pain
12. Is interjurisdictionally coercive, but favorable to alliances.

A boxed summary of all of the characteristics with contrasting poor policy att ributes is presented at the end of the chapter. But let us examine each of them in more detail:

1. Employs Long-term Perspective and Identifies Ultimate Aims

We have talked repeatedly about the need for taking the long view, creating and fostering an evolution toward the very best possible ways of addressing the problem in the long term. This is really a dual element, but the functional part of this policy criterion is the importance of identifying clear aim points. Those have to be as high and distant as possible so that each more immediate step that we take can be seen more clearly as either taking us in the right direction or not. If we build social, economic, and political commitment to a specific energy option that, while offering immediate improvement, can never become truly sustainable on all the sustainability fronts, or will never offer more than a mediocre reduction in emissions, we will have painted ourselves into a corner.

In terms of carbon, specifically, sustainable fuel must achieve near-zero emissions over the course of the next few decades. So, even though we have pointed out repeatedly that we need to start making progress soon, starting off with a strategy that accounts for the need to carry out continuous, ongoing, incremental improvement is still *far more important than delivering immediate results.*

We have to be able to set a path that takes us toward fuel manufacture that is lowest in carbon content, cheapest to produce, and meets the progressively updated, most current criteria of sustainability in other respects. No matter what the policy is, people and organizations will adapt to it and fill the economic niches that it creates. And it is policy death to build constituencies that have interests in fuel options with limited futures; they will not support further change for the better.

Let us think about the virtues of fuel fungibility in that context. Though a necessity, it is also an advantage. A blended fuel product that takes components from different sources, feedstocks, technologies, and commercial interests can be made progressively more and more sustainable, even if manufacturing pathways for certain of those components ultimately prove to be impractical. Our goal can be achieved by policy that helps makers of individual blend components develop more sustainable products, along with policy that helps blenders improve the sustainability of their final fuel by searching out better and cheaper blend components. This is why feedstock-specific or technology-specific mandates, quotas and the like are too shortsighted; they limit overall flexibility in addressing a goal that is manifestly only marginally achievable. We cannot afford to make things harder.

2. Demands Absolute Uniformity of Measure

'Better' or 'cheaper' cannot have meaning unless our understanding of sustainability is uniform and standard. Regardless of any difference in national, regional, or global codes of *sustainability performance*, our methods for measuring or assessing all elements of sustainability—including carbon content—must be identical. Even if the principle of common but differentiated responsibilities forces international air travel policy agreements to consider national preferences, and even if that results in international carriers from different countries being restricted in different degrees as regards their international operations, we have to have a common understanding of what each country, region, and each airline is doing. The same applies to concessions that are route-specific. Furthermore, with regard to publicly funded efforts and private enterprise efforts to produce sustainable fuel *within* a country, policy must compel the measure of fuel products on the same indices of sustainability. On the question of how we measure, the field must be billiard-table-level. On this one point, and regardless of how difficult it is to accept, each air carrier, fuel provider, and national government should agree to a global common sustainability standard measuring system. If we eventually need to have every single air operation on the planet—regardless of where the operating carrier is based, and regardless of whether the flight is international or domestic—using fully sustainable sources of flight energy, that goal cannot be met unless there is global agreement on the criteria and methodology for measuring everyone's progress toward it. National, international, or air carrier standards can develop toward the ideal at different rates over the next few decades, but our way of measuring what is happening in each country, on each airline, on each route, in the domestic or international spheres must be absolutely uniform.

Where some entity or national jurisdiction is driven to take the position that they must have their own certifying body, it is probably acceptable to proceed that way as long as each such body is doing exactly the same thing. For example, both the British DEF-STAN and the American ASTM standards for the technical specifications of fuel are broadly in use, and are equivalent in the matters that users consider essential. Do we need both? Probably not. However, the UK and the US think that we do need both, each for their own reasons. We do not have about two hundred separate national standards because two hundred other countries have decided that the effort is not worth it—and maybe even counter-productive. We should embark on the pursuit of sustainability standards policies with the essential realities in mind: one reason for individual airlines or countries to want their own individual, tailored standards is a reluctance to commit to thorough, rigorous, comprehensive sustainability.

Many international airlines understand the need for a high and common standard; those airlines should find such a standard, apply it to purchases of fuel (regardless of where the fuel is available), make their credible claims about real sustainability, and hope that competitive pressures make other carriers adopt standards that are exactly the same. The major risk is that certain Annex 1 governments will cave to pressure from national commercial enterprises or local jurisdictions. Some alternative fuel makers who cannot meet a high standard will balk. Local jurisdictions with particular vested interests may not be advocates. Such governments (supported by the economic and political weight of objectors) may deliberately ignore the deferred but ultimately much higher cost of having to later recalibrate weak sustainability standards as the need becomes more urgent and apparent with time.

Objections to sustainability standards can also come from quarters where one would not think that it was even an issue. In this regard, we can note the objections being made by the Canadian government and Alberta oil sands producers against the European Union's Fuel Quality Directive, which labels oil sands product as highly polluting; Canada insists that the classification is based on unscientific and incomplete emissions data (Carrington 2013). We will see what the EU decides, but this is (just) one example that argues for applying a standard that assesses *every* source of liquid fuel accurately in all sustainability dimensions.

For obvious reasons, it is in the best interests of those airlines, governments, and fuel makers that recognize the need for high and uniform standards to become more prominent and vocal.

3. Pursues Universal Availability at Uniform Price

The third general policy goal should be that it assists in ensuring that new fuels be available to all, and at the same price. That may not seem possible at first glance, but we will see that it *is* possible, and it will make the alternative fuel option easier to support. Airlines and their home states must not feel disadvantaged by evolving flight energy arrangements. It must be assumed that access to sustainable fuel is likely to become a competitive advantage eventually, and perhaps even in the shorter term. But the introduction of policy related to a broad, necessary infrastructural change should not create winners and losers among those corporations that pay taxes in the national jurisdiction that advances the policy—if they are on board to pursue real progress in sustainability.

Another reason that sustainable, affordable, low-carbon fuel should be available everywhere is that there is no physical reason why it should *not* be, and the policy provision of universal availability might well result in initiatives that see the placement of fuel manufacturing in regions that would benefit from the associated economic development. Fungibility means that regardless of where it is produced and boarded, fuel can be bought and paid for by anyone, anywhere, and bartered for fuels that have to be consumed in some other location. The specific mechanism of proof-of-purchase certification and exchange is perhaps beyond our scope here, but in the abstract, we can contemplate a situation where the first large-scale commercial sustainable fuel production takes place, for example, in a developing nation. As long as there is enough actual local demand for the output of that facility, the people buying the fuel could be located on the other side of the world; they simply buy the fuel, make it available to whoever needs fuel at the local airport, and use the resale contracts (exchange credits) to secure the actual fuel that they burn wherever it is required. In this way, one facility's output could actually be divided up on a proportional share basis among every single airline in the world.

The foregoing examination of the three things that first and most immediately relate to what makes action sustainable sets us up to consider the matter from a more general perspective, and from the approaches that policy development takes.

4. Views Sustainability as an Integrated Concept

Many sustainability-promoting perspectives and frameworks are prone to drift in the direction of putting development first. But we have considered the idea that sustainability subsumes its component 'legs'. Development, for example, will not ultimately *be* development if its mechanisms do not incorporate environmental benignity, social justice, and economic feasibility and advance. But development goals are critical as an element of social justice. So any human enterprise should assist with development goals where feasible and appropriate. Fuel manufacture may include such opportunities.

Beyond development, specifically, every other thing that we have seen as constituting a relevant factor in social justice, environmental, or economic terms must be addressed by policy.

5. Provides the Carrot

This is perhaps the core element of our policy discussion as it evolves. If the development of sustainable alternative fuel (or anything else) is not happening quickly enough, it is simplistic to 'punish it into existence'. We will consider this one point at great length in the next chapter.

6. Complements the Carrot with the Stick

There is good reason for aviation to pursue low-carbon, sustainable fuel, *and to be assisted in that pursuit*. Providing sanction on one side, while also assisting with developing the means of tackling a problem on the other, is a common scheme in broad policy pursuits. The stick must play a role in policies that apply here.

To begin with, there is sympathy for aviation's predicament within the environment community. But it is not unanimous, and it will be costly to lose the support of those who are most likely to then complain vocally and credibly about missteps. A number of environmental activist organizations would like to shut aviation down at any cost. Many of them do not believe a word of what anyone says when biofuels, for example, are offered as a possible part of the solution to aviation's emissions problem. That may be unreasonable, but ignoring the history that has led to this kind of skepticism is not helpful either. Despite all of the advances in aviation's efficiency, total emissions have trended higher. Biofuels technology in other applications has proved to be disappointing; aviation's total absolute emissions reductions and general degree of sustainability would not come near to sufficient levels if it depended upon the sorts of progress achieved with corn ethanol in road transport, for example. All of the tools in the biofuels kit seem to have overpromised and under-delivered.

Nonetheless, there are still important voices within the environmental movement that realize the potential for some kind of sustainable alternative source of fuel for aviation. They also understand that unnecessarily punitive emissions policy may kill aviation's economic capacity for important flight energy innovations. But it seems clear to them (and reasonable to me) that if policy proceeds along the path of providing support to the industry in its efforts to adopt more sustainable sources of flight energy, there must also be coercive oversight pressures brought to bear. And if not immediately and constantly, that pressure should certainly be applied if air transport appears to be failing in its absolute emissions reduction 'endeavours'.

This reality was reflected in an interview with Tim Johnson, who heads up the Aviation Environment Federation (AEF), a UK-based umbrella group with a broad organization membership in the environmental community. In his interaction with a range of environmentalist actors, Johnson encounters a number of positions, including some that regard aviation's efforts to achieve low-carbon flight energy as positive as long as

- other dimensions of sustainability are addressed in a comprehensive way, and
- there is broad policy provision for reducing aviation's emissions to acceptable levels even if the effort to secure sustainable fuel fails. Coercion must be considered an option in that context.

He notes that there is great cynicism within some parts of his constituency concerning both aviation's conviction and the prospects for a truly sustainable aviation fuel. Still, many in his group genuinely support aviation's efforts.

7. Internalizes Effects

As we noted in Chapter 9, an externality is the effect of an activity (negative or positive) on an otherwise uninvolved actor, who did not choose to incur the cost or benefit. Internalization can be characterized as taking responsibility for all effects, intended or not; this ideal is not very compatible with those perspectives that see markets as the only mechanisms that set parameters and commercial interests as perfectly self-regulating. Markets are powerful and valuable and we will undoubtedly rely heavily on their action to make progress on our challenge. But, as we respect their power, we must be equally sensitive to their limitations—they can only transact value where value is perceived; value (effect) has to be internalized *so that* markets can see it and take it into account. Relying on markets to perceive, realize, and clear negative value that is external to them is expecting the impossible.

The key is to recognize negative externalities and use all of the tools at our disposal—science and technology driven by policy *and markets*—to reduce those externalities in a moderate and measured way that is sensitive to developing social, environmental, and economic problems.

8. Establishes Relevance in the Economy

If it is not only important to bring the externalized negatives of incautious technological development and expansion into the economy, they must be properly evaluated. The cost of action should reflect the value of what it is that we are trying to accomplish. In emissions policy to date, this has not been realized: there is no real correlation between the cost of global warming or other climate change effects and the costs of emissions reductions. Is it possible to have cost of damage directly drive expenditures on remediation? Probably not—we are beyond that. The anticipated negative outcomes from the amount of actual or expected carbon in the atmosphere in any likely emissions-reduction scenario are likely already so high that the cost of remediation is necessarily disconnected from what it can accomplish. For example, a comment piece in the journal *Nature* estimated the cost to the global economy of the anticipated release of methane from the seabed of the East Siberian continental shelf at $60 trillion. Whether the release occurs or not, and whatever the exact amount at which its effects would be valued, it is probably fair to say that possible anticipated costs of such events are far outside of our conventional economic experience (Whiteman, Hope and Wadhams 2013).

It would certainly be nice to have an estimate of the cash value of all of the negative effects of our two-century blunder into climatic disruption. For example, what are the economics of sea-level rise? In some scenarios, the scientists tell us that the sea might rise by several meters; even if we are only confronted with a relatively modest one-meter elevation, what will be the cost to the world economy?

Because of the disconnection between the amount of damage and our ability to remedy, these considerations patently do not drive abatement costs. Beyond that, costs of both damage and mitigation are not matters of settled opinion and there may be enough controversy, or at least difference of economic opinion, to block any development of broad consensus about them in the medium term.

But that is not even the kind of question that specific states or sectors have before them anyway. If we are worried that the answer (at this very late date) might be that no abatement amount that we could afford, without bringing the economy to its knees, can prevent the far greater cost of eventual harm, our task becomes that of avoiding as much warming and damage as possible, implying a strategy, at the highest scale, of doing absolutely everything we can do in that effort while still keeping the economy from tripping and collapsing.

Now, if we cannot balance the cost of preventive action against the cost of climate effects due to failure to act, if all of our wealth cannot prevent harm that has an even greater value, the calculation that we are faced with is the relative dollar effectiveness of the things that we *can* do. In other words, our policy initiatives must be appraised on the basis of which ones accomplish the most, if we are committing all that the global economy budget will tolerate. If resolving this seems impossible, it is not. As with other elements of our way forward, incrementalism can work: we can impose *some* value on negative outcomes and tune that value to our capacity to eliminate them. On the other hand, such incrementalism cannot be a simplistic, uninformed, and arbitrary progression. Let's consider that.

The need to focus on longer-term policy aims and to stretch our economy's capacity to react saps our enthusiasm and ignores our general preference for more immediate gratification in terms of both results and political payoff. This deficiency of conviction, commitment, or even awareness, has resulted in policies that lack rationality or suitability. That is far from saying that existing policies are pointless, but they can easily become weak or inappropriate; it is hard to face the dilemma of needing to do absolutely as much as we can and *then* figuring out how much that is.

If the costs of action cannot correlate with the costs of the damage that is being avoided, they simply become the costs of eliminating as much of the source of the problem as possible—carbon, for example. But policy initiatives that penalize emissions generally are forced to apply sanctions that are less than the cost of eliminating carbon emissions (or the economy would halt), and that do not relate to the cost of finding alternative ways to reduce emissions. Though the penalties may be arbitrary, they are applied in the hope that emitters will quickly find the most effective way of minimizing financial pain. But (surprise!) it is not as simple as that.

For example, straight legislative injunction against emissions in the form of limits (to whatever level—50 percent, 0 percent) does not rationally allocate the cost of reducing the carbon emissions. A simple prohibition or cap and trade scheme will certainly produce a set of economic consequences, but the consequences are linked to carbon reduction goals in the abstract, and not to the real costs of doing things in new ways that reduce carbon emissions. A given manufacturer might be thrown into the position of having to choose to stay in business by buying carbon credits before even learning whether there is (or will be) technology that would allow it to continue production output without emissions. In other words, the actual cost of new carbon-reduction technology is only one factor in the mix that determines what a corporation might have to spend. Such considerations would include the state of the economy, the state of its business, the state of other businesses (suppliers, competitors), and the cost of carbon reduction to *those other businesses*.

A market approach would say that policy should provide for a variety of ways for an emitter to take on the challenge of reducing the emissions or compensating for them; it is better to create a dynamic that sees aspiration toward ever less-expensive ways of reducing emissions. Policy must allow for the discovery of sustainable ways of reducing carbon and their introduction into the market at a pace that equals that of the imposition of penalties for failure to reduce carbon. Of course the problem also exists in reverse if there is no financial reason for market uptake and investment in easily available technologies. In the former case, progress can stall, while in the latter, time and money are wasted as the carbon problem is allowed to grow more quickly than it otherwise might have.

9. Discloses Costs

In the case of cap and trade schemes, what an emitter will pay for carbon credits does have something to do with what they would otherwise pay to implement carbon-reduction technologies, but the value of those credits is determined by political decisions: how many credits to grant for free, how rapidly to lower the cap, how much demand for credits is driven by the state of the economy. If, in an evolved cap and trade scheme, carbon is trading at $20 per tonne on a given day, this reflects the market value of the carbon avoidance driven by:

- arbitrary limits on emissions
- arbitrary amounts of emissions credits initially made available
- arbitrary initial costs of initial allotments
- costs of the various ways of reducing emissions
- the varying amount of carbon-generating activity, and the varying amount of carbon-reduction technology in the economy, *taken together*.

In the case of taxation schemes, the rate is set according to straight political calculus and does not necessarily correlate very well to the cost to the emitter of reducing their emissions. Policy must not only be effective, but it must be rational. In fact, if it is *not* rational, it cannot really *be* effective. It is rationality that allows transparency.

10. Directs Revenues Toward Solutions

If the policy goal is to urge as much action on emissions reduction as possible without crippling the economy (which has not been the case so far), policy-makers contemplating sanctions are in the position of having to decide how much wealth-producing activity to take out of the economy and where to put it back into the economy, while realizing that wealth production can also act to fund alternative ways of doing things. So one would hope to see revenue from such a mechanism flow toward assisting in the development of alternatives. In fact we absolutely *need* to see that. The provisions of the European Union Emissions Trading System (EU ETS), as an example of policy in place, make no such commitment; any funds generated are allocated to the states' general revenues. In turn, general revenues are expended in a largely carbon-intensive economy. This is like bailing water out of one part of the boat and pouring it back in another.

11. Provides Some Immediate Benefits with Minimization of Pain

Although we must start slowly and cautiously, we must respect the need to make tangible progress quickly. Simply getting started (at any level of action) is hugely important. But as human societies, we are chronically, devotedly, and perhaps (in this context) pathologically reactive. The political challenge is to show how the first moves can bring more immediate benefit than sacrifice. And our first priority must be that we *do not make the situation worse*: a sort of political, economic, commercial, and social Hippocratic oath. In that regard, recall our first consideration, *Employs Long-term Perspective and Identifies Ultimate Aims.* Get started right away in a way that recognizes—*clearly*—where we have to end up.

12. Is Interjurisdictionally Coercive, but Favorable to Alliances

On the political front, there is sometimes no real or practical acknowledgement that there is urgency. Society has not completely absorbed the reality of the enormous costs of global warming. And politicians, needing popular support and fearing outrage, are often reluctant to make the situation perfectly clear. Even worse, some ally themselves with those who deny the climate reality for short-term economic gain. And even those political leaders who might otherwise support sustainability efforts must contend with other jurisdictions that can gain at least short-term advantage by refusing to adopt similar stances. As a consequence, levels of commitment to abatement are arbitrary and nothing like as high as possible. Certainly, one way to address that situation is to craft policy in a way that exploits the power of international economic and diplomatic levers. Even more powerfully, sustainability and policies in economy, trade, and diplomacy can be amplified through pooled commitment.

Armed with descriptions of the characteristics that we would want policy to reflect, it is now time to see who the appropriate actors might be, and how they might act to create and implement policies that fit the bill. A summary of what has been discussed is given in Figure 11.3.

Desirable Policy Attributes	Undesirable
o Employs long-term perspective and identifies ultimate aims	o Shorter term focus and strategies do not align with most distant aims
o Demands absolute uniformity of measure	o Allows development of multiple standards systems
o Pursues universal availability at uniform price	o Focuses on local needs
o Views sustainability as an integrated concept	o Sees elements of sustainability as distinct targets
o Provides the carrot	o Lacks positive incentive
o Complements the carrot with the stick	o Lacks element of sanction
o Internalizes effects	o Externalities unrecognized
o Establishes relevance in the economy	o Costs of compliance improperly related to goals
o Discloses costs	o Costs unclear
o Directs revenues toward solutions	o Revenues flow to larger economy
o Provides some immediate benefits with minimization of pain	o No immediate benefit for policy targets
o Is interjurisdictionally coercive but favorable to alliances	o Isolationist in scope and conceptualization

Figure 11.3   Summary of policy characteristics

Chapter 12

The Machinery of Policy Development

How does policy get made? How do people think about policy? If we now have some ideas about what makes policy good in terms of helping to achieve an industry's goals, what do we need to know about policy development that would help make some good policy approaches come about? What follows is, essentially, a contemplation of the policy factors that bear on the aviation industry, a comment on some existing policy initiatives, and a few tentative ideas about what might be proposed, whether additional to or in place of other policy.

The Personality of the Actors

It is never enough to only know what a policy is meant to accomplish. The first step in creating a policy—any policy—is knowing the perspectives, priorities, and modes of operation of the entities that generate it: political establishments and policy establishments. National governments hold the capacity to act, but there are decisions to make about how they should act and how to move them to act, and those decisions have to be made by everyone who has an interest in a policy goal. When it comes to broad approaches to policy, governments have 'personalities' that inform their larger view. The actions and approaches of various ministries and departments may be driven in their respective policy development exercises by official meta-policies, or by implicitly understood though ill-defined philosophies; nonetheless, these personalities exist, and are powerful and pervasive. No matter how policy approaches vary, and regardless of their priority in a state's political evolution, people and their governments look at policy questions from a particular point of view and adopt a particular policy orientation. For those of us who support the general thesis of the significance of sustainability and how it operates, it is important to understand such orientation, work to influence it, and operate within those results. For those actually engaged in making policy proposals, or seeing proposals through to realization, understanding policy orientation is critical. Efforts at policy advancement must be pursued with an awareness of where interests and goals might intersect or potentially operate at cross-purposes. When policy development seems to be frustrated, arguments *for* the policy, and for any implied need for an adjustment in general policy orientation can then be made cogently and in a way that is least disruptive to political goals.

Though some policy ideas will be mentioned, it is beyond the scope of this book to be prescriptive of strategies for addressing or producing effective adjustment of policy orientation. But it is necessary for those involved to acknowledge the what and why of policy orientation and its implications. If we are fortunate enough in a given circumstance that we enjoy orientation that might favor an evolution toward policies that would support sustainability in general, or sustainable aviation fuel in particular, we are lucky. If not, it becomes necessary to formulate policy ideas in a way that best conforms with the perspectives of those who must enact it. I mention orientation, ideology and the like principally to distinguish them and set them apart from what we will discuss next.

Policy Theory: Ecological Modernization

Assuming that the political establishment accepts the idea that policy has a role to play, and that we are prepared to view the matter within the parameters of *whatever* policy orientation exists at elected-government level, we are somewhat at liberty and equipped to make general suggestions. In doing so, it is necessary to find an approach to policy development that might appeal to anyone, regardless of the orientation about which we just spoke. We must seek a prevalent, accepted, and useful framework or general, theoretical way of looking at the job of policy development. This does not relate to larger political or ideological goals, it is rather the nuts and bolts of how politicians and bureaucrats view the logic of policy machinery. If we can engender or expect an orientation that favors or accepts the goals of sustainability, what sort of logical formulations of such policy approaches are readily available? What policy development toolkit do we pull out of the locker in order to pursue our goals within the prevalent orientation framework?

There are a few different schools of policy thought when it comes to sustainability. One that gains currency as a perceptual frame and policy path for sustainability initiatives is ecological modernization (EM). EM seems to be the framework that attracts the most attention, though some rather (in my view) misguided understandings do exist, and they should not be proffered as remedies. The application of EM is illustrative of how the misapplication of a theory or standpoint can set back an agenda.

EM theory can be quite useful. In brief, it implies that policy and technology represent the practical and essential tools that can be brought to bear in addressing the need to be environmentally responsible. However, EM theory can be misinterpreted to imply that unlimited growth is possible, and the role of policy, in that context, is to actively foster a straightforward technological assault on anything that stands in the way of economic expansion, while ignoring a myriad of relevant underlying issues, priorities, and drivers. Some attempts to develop sustainable alternative sources of flight energy are criticized as pandering to the worst of such misconceptions. In fact, EM theory *does* argue for the capacity of government, policy, markets, science, and technology to address the environmental issues that confront us, but it does not necessarily imply uninformed assumptions about side effects. Critics argue that EM perspectives assume that every environmental problem must have a policy and technological solution that encourages economic growth without condition, and that arguing the opposite is either naïve or presumptuous.

Probably a little bit of each respective view is valid. It does seem simplistic to think that we can continue to make our world's societies and economies larger and more complex without recognizing that this expansion generates a new and uncomprehended version of that world. So criticism of EM strategies is spot-on when those strategies imply that there is always a technologically rational way of moving forward.

One way of looking at this is that expansion, in both the kinds of newly innovated technologies together with the growing magnitude and scope of their application, tends to create effects in a non-linear way; unintended effects increase in number geometrically, and technology interactions create unanticipated discontinuities. In other words, this relentlessly expansionist expectation really constitutes a presumption about a future that we cannot, as a consequence, envision very clearly. Everything is a complex interaction of technologically enabled effects that are growing in absolute size, scope, and number, with an ever-wider range of characteristics. Even if we put a 'green' emphasis on everything and try to focus on sustainability, we are still enlarging our economy and its effects in both numbers of technological features and outcomes as well as absolute size. All of this raises an important question with policy implications even if we are focused on improving sustainability: If the results of our actions become increasingly harder to understand and predict, how can we know that continuing those actions will lead to a desired outcome (especially when things seem to have gone very wrong)?

The answer to that question seems tricky. Probably the answer is that we cannot rely on doing more without also understanding that action creates risks—and then figuring out how to manage those risks. At first blush, this seems like a weak answer. But, fair or not, proponents of EM have to accept it.

On the extreme other hand, wholesale rejection of any possibility of dealing with our problems through innovation in policy and technology is not a useful perspective either. Some of EM's optimistic assumptions make it attractive as a way of envisioning how society can advance and how policy should support that advancement. We do have to manage our way out of our difficulties, and simply shutting everything down—while certainly arresting many of the things that we are doing wrong—creates its own set of unknowables.

The irony is that our economy and its technology are the only tools at our disposal. Our society is the essential, defining, total product of our species as a social animal; it and its resources are all that we have, both as starting point and apparatus to continue to act. We can only act in a way, and to an extent, that our societal framework or establishment can conceive and enable. Since our ability to act is a socially framed capacity, we need to conserve the integrity of that ability by adjusting rather than destroying its configuration. Precipitating an abrupt change can create a revolutionary discontinuity or inflection, and can compromise our ability to continue to act. For example, sudden constraints on all of the things that enable current economic and social arrangement could result in the migration of power and authority into hands that care even less about the environment, society, and economy. This is the very story of revolution gone wrong in some other contexts. So, simple constraint on harmful activities has its limits as a policy option, and the constraints must be crafted carefully. We must change our ways in a manner that accepts the need to both maintain our economic power and vibrancy, and also to remain cognizant that an economy that is developing in both size and complexity constantly produces larger and more complex unanticipated outcomes, and that keeping track of, understanding, and controlling for these outcomes is at least as important as ensuring that the economy, in a viable state, is husbanded along.

In that sense, it is probably fair to say that EM, *if it is viewed simplistically as a panacea* (important qualification), is dead in the water as a useful framework. But something like EM is essential, as a vital way of understanding and moving toward a more sustainable world, because we must apply policy and technological approaches toward addressing our problems. However, analytical tools and policy theory frameworks (and these are the roles that EM seems to serve) must shift emphasis to real sustainability. Technological and policy fixes to environmental and other problems must advance in a way that looks as far forward as possible, keeping possible indirect effects of the multiple dimensions of sustainability at a world scale firmly in view. Consequently, they also imply incrementalism and measure.

Christoff (1996) proposed the term 'strong EM' to describe a version that is diversified, deliberative, and comprehensive, and that focuses on the need to be part of a solution rather than just less of a problem. 'Weak EM', then, becomes the narrower, more undeveloped version, with an ill-advised and perhaps somewhat blind dependence on technology to come up with answers. People have taken up this EM strong/weak formulation and the theory is now often discussed in those terms, not only in understanding policy options, but also in assessing past actions and initiatives already in process. In that light, if EM is to be taken as a useful policy development lens or filter, it must be adopted in its very strongest 'Strong EM' form.

Though the foregoing is a very brief description of just one theoretical framework, it is the one that I think that policy developers are most likely to be familiar with and to find helpful in presenting ideas about elements of sustainability that might fall within their jurisdictions. It is a way to view policy development in terms that are both logical and encouraging, and, if applied properly, may even help to overcome a political (ideological) predisposition to reject government action as inherently undesirable.

Policy Development: Whose Responsibility?

The question of who can and should act is also critical. A decision to compel reductions in emissions requires effective unanimity at some level. Where unanimity is lacking—as in the case of the intermittent desire of the European

Union (EU) to impose its regional Emissions Trading System (EU ETS) on international aviation, and the world's willingness to accept it—such decisions provoke disagreements that threaten to lead to trade sanctions and a general descent into confrontation. Where does the capacity reside to encourage greater use of technologies that *will* reduce emissions?

Though the world may be slow or reluctant to find ways of acting together to force reductions, acting more locally to bring emissions down is not the same kind of problem. Decisions to simply support ways of providing technology resources that can reduce emissions are not so fraught. Individual carriers, states, and fuel providers can get busy on that, and there are a number of reasons why they should. So we should remember this general principle: the more local the jurisdiction and the participants, the more reliable is the ability to act on policy matters. Parenthetically, it is at the level of the national government where authority is exercised over lower jurisdictions and where positions to international bodies are decided. But it is in the hands of enterprise where the most powerful *action* tools are held. Therefore, one view of the particular kind of policy task that the sustainable flight energy challenge conjures is intensely integrative: using the direct connection between corporations and local governments that are able to see complementary interests, and taking resolve that is built jointly at that level to motivate national governments, since they can do two things: scale to make the idea nationally advantageous, perhaps realizing better economics at that scale, and promote the basis of the idea to international relationships.

Since policy at the international level flows from decisions made by state governments, the policy matrix that we are discussing involves both domestic and international geographic scope. It also includes both incentive (carrot) and deterrent (stick) approaches. It becomes difficult if not impossible to keep these four dimensions apart and talk about them separately. So our discussion is comprehensively about how states can help the development of sustainable fuels, how they can act collectively toward that goal, how they can participate in spreading new fuel technologies into all parts of the world, and whether they can collectively or individually compel charges on the aviation industry where it fails to attain the goal. But even if it requires a level of detail and complexity that cannot be entertained here, remembering that policy action and goal-setting by more local or regional governments together with corporations is absolutely essential; this is the 'kindling' that forces national governments to engage.

Review of the Interplay of National-International-Carrot-Stick

The International Civil Aviation Organization (ICAO) acts as the aviation body that sets forth the collective will of states (where collective will exists) and is structured to be able to formalize that will. States decide what happens at ICAO. The matter of aviation GHG emissions fits comfortably within the ICAO ambit. Could the ICAO forum be used to establish limits on GHGs from international aviation and establish a system of sanctions for failing to achieve specified reductions? Perhaps. But it is unlikely, because there would almost certainly be some states that objected and would simply file an intent to ignore the agreement—which is their right. Alternatively, states could perhaps agree to have ICAO standards and recommended practices (SARPs) spell out the ways in which aviation's GHGs will be accounted and compensated for in some way. This is the possibility that preoccupies ICAO's community. Without having to actually force reductions in emissions, ICAO delegates *can* establish a system of simply and impartially forcing a cost on carriers for the emission of carbon. This would presumably be achieved through the application of MBMs. Such discussions are ongoing. This is a lower hurdle than compelling emissions reductions but the countries cannot seem to agree on how even this straightforward exercise should be done. A point (important to us, here) that becomes clear as a result of this paralysis is that it is unlikely that any ICAO agreement could see the compulsory adoption of sustainable alternative fuels. Countries want the latitude to decide how they will reduce emissions; MBMs are one option that represents a very limited, straightforward way to get started on the emissions challenge if agreement is possible.

In the United Nations Framework Convention on Climate Change (UNFCCC), states have agreed that dangerous global warming must be averted but have not (except in the Kyoto Protocol) agreed how much should be done or by which nations. However, since we have seen that there is no *dis*agreement that international aviation will fall outside of any individual national commitments on GHG reductions (should they ever be defined in a post-Kyoto policy environment), states should be agreeable to having an ICAO formula address the problem. But, in the end, some countries do not want this, some want to be afforded special status, others see complete breakdown of the commercial configuration if special status is granted, and still others cannot agree to accommodate either of those two perspectives. So we could say that in the 'international-stick' sector of our discussions, things are badly stalled.

Can there be an 'international-carrot' program? If the stick means punishing emissions (or making everyone account for them—same thing), a 'carrot' has to be helping the industry to get rid of the emissions so that there is no need of punishment. That means helping with efficiency improvement and reduction of carbon in flight energy, and only the latter takes us to the ultimate goal. ICAO has no resources of its own to positively incentivize the propagation of lower carbon fuels. And for reasons that would be similar to those that have stalled MBMs, it is hard to see states agreeing to whatever charges would be necessary to fund such incentives. (In fairness, it must be pointed out that the ICAO Secretariat has been very active in bringing states together to discuss the importance of fuel options. And I can say that such discussions have been, in my opinion and by the accounts of most with whom I have spoken, fruitful.)

It is probably best to assume that our *best* ways forward reside in each respective country's legal capacity to:

- forbid or limit action (including emitting noxious substances)
- require action (such as the procurement of sustainable fuels or carbon permits *in lieu*)
- create the incentives or assistive mechanisms that would foster the desired action or result.

This leaves the actual exercise of reducing aviation's global warming wingprint, or creating the capacity to do it, in the hands of airlines, airline associations, and corporations that offer new flight energy products, along with national, regional, and local governments who regulate their corporations and control the policy arsenal. If we can just register those general facts, we can embark on the broader discussion.

The foregoing summary is useful as an introduction, I hope. But we have not come this far to dismiss policy matters on the basis of a few paragraphs of generalization.

ICAO as Facilitator

We should probably try to understand the limitations that inhere in ICAO's processes a little bit better so that we really know where the barriers and the portals are in this whole matter. Let us recall a little more specifically what has gone on. If international aviation (as opposed to domestic) is the place to *start* in looking at carbon, fuel sustainability, and the usefulness of the ICAO forum in these matters, it is also necessary to accept that international agreement tends to settle on the most general goals and steer away from the hard requirements. Resolution A37-19, adopted at ICAO's 37th session of the Assembly in 2010, says that states and relevant organizations, working through ICAO, would commit to achieving an average 2 percent per year improvement in fuel efficiency to 2020. After 2020, the aspirational goal was to continue to make 2 percent per year improvements in fuel efficiency, and to prevent global net carbon emissions from international aviation from exceeding the 2020 level (International Civil Aviation Organization 2010). We can argue about whether these hoped-for levels of achievement are sufficient to the challenge, but their biggest shortfall is their lack of prescriptive effect. Further, they evade a needed commitment to reduction in emissions in absolute terms, favoring instead a focus on energy intensity of flight and then a cap. But we need reductions, not just stasis.

ICAO Delegates' Policy Strategy: Market-based Measures

Market-based measures should almost certainly play a role. They comprise a menu of arrangements from which it should be easiest to agree upon a particular selection. The prominence of their role will depend upon the success of new aircraft, operational procedures, airspace management (ATM) improvements, and—of course—new fuel sources. But no matter how effective any of that is,

MBMs could conveniently fill gaps. The following graph shows how MBMs will probably be required in industry emissions reductions for at least an interim period. We have talked about goals of carbon-neutral growth (CNG) as of 2020 and a 50 percent reduction by 2050. The notional curves show approximations of growth in aviation emissions moderated by technologies and efficiencies. We can see how the introduction of new fuels is relied upon to accomplish everything else. But new fuels will take time, and so there is an area under the new fuel line and above the CNG line where MBMs (or 'economic measures') should be the answer.

I have seen a slide similar to this (often from more than one presenter) at virtually every sustainable-aviation event that I have attended in the past several years.[1]

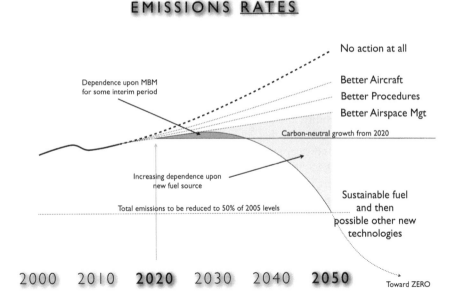

Figure 12.1   Role of various factors in achieving emissions reductions

1   For examples, see the graphs in the position paper of the Air Transport Action Group (ATAG): http://www.atag.org/component/downloads/downloads/230.html. Airbus commitment statement: http://www.airbus-fyi.com/article/215?back=Site.about P. Steele (International Air Transport Association) presentation: http://sustainabledevelopment. un.org/content/documents/PaulSteele.pdf

Like bank advertisements about saving for retirement, the schematic makes another point: the earlier one starts, the easier the overall implementation is and the less drastic the implementation measures must become as time passes. Getting any significant (even if small) amount of sustainable fuel production capacity in place as soon as possible changes the shape of the curve substantially, and consequently changes the MBM area under the curve as well. However, since MBMs preoccupy the international policy forefront at this time of writing, we will continue with them and describe the discourse, because the industry considers it critical to be able to offset emissions with credits as required for some time.

The MBM Discussion

It should be noted that, aside from the immediate help that MBMs could offer, if MBMs and accompanying limits were to win through to implementation, a lot more people might be motivated to solve the fuel problem in order to avoid real costs, even if they thought that CNG-by-2020 was irrelevant. If nothing so focuses the mind as the prospect of a nasty outcome, it may be that an agreement to implement MBMs will bring about a certain amount of concentration.

Among MBMs, the mechanism that enjoys most support within the airline community is offset purchase. But the choice of specific MBM mechanism to be pursued is not as important as the inherent barriers to *any* mechanism. The matter that makes the discussions difficult is our old Scylla and Charybdis: the Chicago Convention's uniform treatment of international aviation carriers on the one hand, versus Common but Differentiated Responsibilities (CBDR) for states on the other. Is there a way to navigate between imposing MBMs on everyone, or losing the whole idea in the maelstrom of trying to rein in only certain carriers and countries?

Let us start off by slightly elaborating on the point about the relevance (or lack thereof) of national boundaries in some circumstances. Air travel into any country is extremely valuable. It is not the country itself that produces flight emissions, it is the flying, and by implication the people who are doing that flying. Obviously and essentially, less-developed countries will want to avoid any kind of barrier to increasing the number of people traveling to their cities. But if aviation's emissions are to be addressed, they must be attributed. Any person who has the wherewithal to fly should account for their emissions—*every* flying person, whether or not they travel *to* or live *in* a developing country. Though the official discussion revolves around the obligations of sovereign states, states do not make carbon dioxide. It is necessary to divorce international air travel emissions from geography and tie them to the actual activity and the people who indulge, otherwise the momentum of current talks that will inform any first attempts at an international flight MBM regime will spend itself figuring out ways to let certain (generally disadvantaged) countries off the hook. Allowing whole countries to escape their responsibilities for international flight emissions has the effect of allowing many individuals (who could afford to pay) to escape *their* responsibilities. It also raises the specter of trade sanctions and the opening of the relevant bilateral agreements.

Right now, shorter-term policy efforts on the international front at ICAO advocate for releasing certain international flights from their carbon-accounting obligations on a route-specific basis, where such routes serve a qualifying country. Perhaps no airline would have to account for its emissions on such routes. Carriers based in relatively wealthy countries would be accountable only on routes that did not go to poorer countries, whereas airlines based in poorer countries would be accountable for nothing. This may prove to be an effective way of addressing the concerns of countries that want CBDR relief, while treating every airline on such a route equally. It is, in many respects, ingenious, elegant, and attractive. And, by the way, it is not the only possible one; the air industry organizations have a different idea, and we will see a bit more on that later.

Before we go further, it should be understood that *any* such relief must be offered solely in the context of a policy pathway that evolves toward completely comprehensive carbon accounting for every flight on every route, *eventually*. Remember that the totality of international aviation must achieve the change to *very* low (approaching zero) carbon emissions from fuel burn, which cannot happen if any substantial amount of flying is allowed to continue with fossil fuel. Any policy provision that caters for special considerations—whether targeting carriers or routes—would have to contain sunset provisions that would see their effect diminished over time and become virtually null in the years approaching 2050. In that regard, again, identifying flying activity as 'individuals moving themselves around' allows us to consider things in more neutral terms. Just as with other activities that constitute 'being developed,' or 'wealthy', every individual will have to be performing those activities in ways that are supported with properties of environmental benignity and social justice.

Divided Opinion at ICAO

Now, remembering that the industry actually *wants* international agreement on MBMs and also that circumstances dictate that they can only be a certain amount of help for some limited amount of time, what prospect is there for agreement on MBMs? I am struck by the impression left with me in the aftermath of the 38th Assembly of ICAO, 2013 (the triennial meeting of ICAO's supreme body, in which all Member States may participate). The Report of the Executive Committee on Agenda Item 17 (Section on Climate Change) includes a general agreement that MBMs are potentially useful emissions-reduction strategies, but also reflects a lack of consensus on which measures should be adopted, by whom, in what way, and to what purpose. The only real conclusion was that more study and discussion were needed, and that in all circumstances MBMs must not impose an inappropriate economic burden on international aviation (International Civil Aviation Organization 2013d, International Civil Aviation Organization Executive Committee 2013).

It would be irresponsible to simply call the 2013 Assembly a failure. These things are never that simple. Still it has to be acknowledged that many have been disappointed by the tenor and fruit of ICAO discussions leading up to the event, together with the rather lightweight Assembly Resolution A38-18 on Climate change, various sections of which generated reservations from individual countries (International Civil Aviation Organization 2013e). Another thing to note is that the failure does not belong to ICAO itself. It is the states and their delegations that have failed.

In discussion with Chris Lyle, who represents the World Tourism Organization at ICAO and has a long career in aviation including many years at ICAO itself and as an independent consultant, it became apparent that there was really little to report in terms of progress in squaring the policy circle on global MBM for international aviation. He follows ICAO discussions on sustainable aviation as closely as many, and is a particularly astute observer and commentator.[2] In the context of some degree of my prior hopefulness, he described to me what he called the 'sequence of unraveling' that took place in Montreal. In Lyle's view, Resolution A38-18 is generally a weak document, on the modest provisions of which many countries have lodged reservations (see the Annex to the Resolution). Particularly notable, in the context of a position taken by the international airline industry itself, that it is necessary and possible to achieve carbon-neutral growth by 2020 and a 50 percent reduction in absolute levels of emissions by 2050, it is astounding that this body—which comprises representation from all of the governments where the airlines are actually based—states in Paragraph 7 of the Resolution itself that CNG by 2020 is an 'aspirational goal' but *not* an absolute requirement at all. A number of factors are mentioned, but the resolution's summative phrase explains, 'emissions may increase due to the expected growth in international air traffic until lower emitting technologies and fuels and other mitigating measures are developed and deployed' (International Civil Aviation Organization 2013e). This is patently contradictory inasmuch as the whole discussion about CNG has been mounted on the basis that such commitments should be met *notwithstanding* growth. In fact, the resolution begs the question, implying that while carbon-neutral growth is the goal, it may be impossible to achieve because there *will* be growth and it may *not* be carbon-neutral. This ignores the whole point, by ignoring all of the ways and means that the Assembly is considering in order to *achieve* carbon reductions (including MBM, which most participants regard as the specific tool to be used for the express purpose of achieving carbon neutrality until better fuels and other technologies are available).

Anyone who calls the document in question a complete nonsense could be forgiven. And characterizations of the session as a 'landmark' and the agreement

---

2 For a very useful summary of events as they led up to the ICAO Assembly, see Chris Lyle's articles at Green Air Online, http://www.greenaironline.com/news. php?keyword=Chris-Lyle%2C-ICAO%2C-UNFCCC%2C-AGD-Group%2C-AEA%2C-AGD%2C-Aviation-Global-Deal

'dramatic' (International Civil Aviation Organization 2013b) do seem overstated. But such an assessment would lack any subtle understanding of how easy it is for nearly two hundred countries to tie themselves in knots when the issues before them are complex and challenge their respective interests in hugely disparate ways. And there has, in fact, been useful progress in some particular areas.

But the ICAO resolution does fall short. And while appreciating the difficulty of the process, we still need to be very clear about how little it seems likely able to achieve. Essentially, the 'dramatic' agreement is that MBMs do offer potential and promise, require more study and discussion, and should be brought up again at the next (39th) ICAO Assembly in 2016. Many of the states' delegates certainly will have worked to prepare a proposal for consideration at that time, and it is hoped that an MBM scheme could be approved then for implementation by 2020. Under the circumstances, it would be a tad ingenuous to rely on even that prospect, but there is a common awareness that coming to some kind of agreement is necessary, so we should certainly hope and encourage.

An example of why such agreement is important to industry is the outstanding (at this date) issue of the EU ETS. In a February 2013 speech, discussing the temporary but crucial need to implement some form of market-based measures in order to deal with emissions reductions, the Director General and CEO of the International Air Transport Association (IATA), Tony Tyler, urged the airline industry to accept compromises as policies are formulated, and warned of the consequences of failure to present a united front:

> Recognizing that the process for governments to reach an agreement will be difficult, it is in the industry's interest to provide as much support as possible. … The incredibly complicated and burdensome monitoring, reporting and verification (MRV) requirements of the EU ETS proposals are a clear example of how things can go terribly wrong when we leave it to governments to decide how we should run our businesses. (Tyler 2013)

CBDR as a Complication

Events subsequent to the 38th ICAO Assembly are no more encouraging. In news reports, the BRIC countries (Brazil, Russia, India, and China) jointly propose that a new group, the Environment Advisory Group (EAG) be put together for the purposes of figuring out what should happen next and at the 39th Assembly in 2016 (GreenAir Online.com 2013). They announce that their proposals would specifically entrench CBDR in any MBM scheme. The draft proposal also apparently insists that countries and regions absolutely refrain from unilaterally imposing MBMs on international flight. This would mean no action until there was general international agreement, and that general agreement must incorporate CBDR. Well-intentioned though it may be, there is just too much there that would strain the will of countries to agree or, alternatively, limit the effectiveness of a proposal.

What is the fallout of such paralysis? If no agreement is reached in the ICAO forum, the whole challenge—sticks, carrots, interests, and obligations all in a vast compost heap—is consigned to the hands of governments, air carriers, and fuel providers, any of whom might be prepared, each for their own reasons (whatever those reasons are), to produce solutions. They are the ones who are working on the ultimate solution to the carbon problem as opposed to being preoccupied with a temporary post-2020 need to be able to buy carbon credit.

However, although they may be more likely to create solutions, they may be less likely to share them in the aftermath of the MBM/CBDR fuss. That is a matter of critical importance because any such disregard would defeat the will of our larger institutions of international scope, whose salient intention is to get carbon-reducing technologies into the hands of those who otherwise lack the resources to develop them. It cannot be allowed to happen. Out of the mix of realm, interest, and scale as among communities, regions, states, and corporations must come a way of resolving a matter that greater international institutions (the UN and its conventions themselves) have determined *must* be resolved. We all know that it must be resolved. When the lesser body with specific responsibility for international air transport cannot get started, we still need low-carbon fuel and we need it in everyone's hands, not just the companies or countries that undertake to develop it. Regardless of what happens at ICAO with the MBM battle, it cannot be allowed to paralyze everything. Though MBMs cannot ultimately deliver actual carbon reductions, the fact that they will likely still be required is important, and fighting over how they should be applied, whom they should affect, and how they should be implemented can create fissures that do not heal themselves when we need to deploy new fuel and other technologies into all of the parts of the world. That matter needs some thought.

In any event, and to sum up the situation: market-based measures, in some form, are the most suitable and reasonable action to take in applying any sort of 'stick' to aviation emissions. But application of such measures in the international arena is problematic and delayed, and their application in narrower jurisdictions is even more difficult.

The air industry itself regards MBMs as essential in the period after 2020 when the commitment is to industry growth that will not raise emissions. And the industry feels that the need for MBMs will continue until such time as combinations of technologies, including sustainable fuels, are able to bring actual emissions down to targeted levels. We are used to situations where governments impose sanctions and industry bears the brunt. In the current situation, industry embraces the brunt, and governments refuse to work together to impose it. Some individual jurisdictions *would* impose it, but their neighbors refuse to cooperate until a joint system is created—to which they would then refuse to agree. (Boggle at will; you are not alone.)

A further thing to note about the disagreement over MBMs: Many of those who favor MBMs would use them to extract revenue from aviation that would not be put toward the purpose of solving aviation's emissions problems. They would penalize commercial flight for emitting GHGs, and deposit the penalty cash back into state general revenues to be spent in a persisting carbon economy. It becomes not only unfair, but also pointless and even counterproductive.

Are MBMs Stealing the Spotlight?

All of this points out how critical it is to broaden the discussion before the will is completely compromised. In what should be a four-dimensional 'national-international-carrot-stick' matrix, MBMs are the international 'stick' that is using up all of the policy-generation energy at this time. And it is not clear that they could be applied fairly, even if CBDR could be addressed. Not all of the possible effects of applying sanctions have been thoroughly considered. The imposition of costs affects individual market actors differently: each player has its own limits on its ability to absorb costs, and MBMs would create some degree of commercial perturbation.

Of course, each airline, or airline paired with its respective home state, also has a different capacity to introduce low-carbon aircraft, procedures, and fuel, but there we may be able to call upon other policy drivers to ease the differences—if we do not argue so aggressively about MBMs that we cannot talk about anything else *to* anyone else.

Possibilities of Unilateral Action

We should just touch upon the following before we leave the subject: As has been mentioned many times in different contexts, the EU, refusing to be hamstrung by failure in international forums, has started to experiment with its own MBM policy options. This action is worth exploring, because a contagious outbreak of MBMs is perhaps as good as an agreed imposition.

One question that arises concerns geographic scope: should a country or federation be allowed to act beyond its borders? The natural and widespread view is that it should not. But regardless of popular opinion and court rulings, it is probably unwise to dismiss the idea. Countries certainly *can* act outside their borders. Actions, no matter where taken, can sometimes be felt in other places, and the people in the affected places may create laws with extraterritorial application in order to address those impacts.

The EU ETS is an interesting case. In one incarnation, it imposed conditions on all flights into and out of EU airports, regardless of whether or not the carriers' countries are members of the European Union. Hertogen (2012) provides an excellent discussion of the jurisdictional issue, and concludes that actions such as the EU ETS at least provide impetus for cooperation, incentives for states to harmonize their efforts, and some form of uniform control. 'However, we need to question whether, while we pursue the holy grail of an effective and binding multilateral agreement, we prefer no regulation or unilateral regulation' (Hertogen 2012).

Since EU action might precipitate similar actions in other jurisdictions, such unilateral measures could perhaps be a desirable way of enabling the necessary (if stopgap) imposition of MBMs. Would such a plan be contagious? We do not know. But if it does not spread, and become universal quickly, it can easily make things worse rather than better by killing the still-necessary will to work together on other ways of addressing carbon even if imposition of MBMs is stillborn.

Revenues and Exemptions

One major flaw in the versions of EU ETS implementation that we have seen has to do with the allocation of revenues. It has been proposed again and again by states' delegates to the European Parliament that cash from the EU ETS auctions be distributed to member states with no legally binding provisions for how the money should be spent. We have pointed out the error in this approach but states are vehement and disregard the effect of allowing revenue from a carbon-reduction plan to flow, un-earmarked, into the general revenues of countries whose economies are still largely carbon economies.

If the decision to apply the EU ETS to international aviation stands, nothing prevents other jurisdictions from establishing their own cap and trade schemes and selling all of their allocations to raise general revenue. Such an eventuality entirely defeats the purpose of the supposed punish-to-improve logic. Aviation emissions might drop, but the money exchanged may very well flow to other emission-stimulating activities in each of those economies.

If we think about Hertogen's analysis and the logic of cap and trade, it is in the air industry's interest to be very vocal about the need for public information concerning the application of revenue that accrues from these schemes. It is essential for everyone to understand the details because (for example) the extension of the EU ETS to international aviation is being sold on the basis of overall carbon reduction. The real effect cannot be known unless we can track the revenue. This is really a hill upon which to die.

Just Reduce the Carbon

If turning the key will not start the car, we have to consider whether it might be better to just get the thing rolling, put it in gear and let in the clutch. The preoccupation with the MBM debate becomes an exercise in futile anger for some, and a way of delaying action for others. It is logical and crucial for the industry to be aggressive in figuring out the low-carbon fuel puzzle, the only way out of a potentially debilitating, destabilizing morass. Carriers who may be made participants in a cap and trade scheme or even an offset purchase scheme are better served by avoiding it (or participating as little as possible) in favor of being able to offer real carbon reductions that have validity in any scheme.

Note that we talk about the interests of airlines that *would* be subject to the provisions of MBMs. These carriers should act to get low-carbon fuel into their operations regardless of any exemptions that other carriers enjoy. It is with the discussion on policy for the support of new sources of fuel that we will carry on in the next chapter.

Chapter 13
Policy Strategy: Support

Our policy discussion, thus far, points in a definite direction: if policy initiatives in the international forum are ponderous and slow, where are the opportunities to be helpful more quickly? Specifically, if finding new sources of fuel is the answer, what helps that happen? Agreements reached through the International Civil Aviation Organization (ICAO) and international compacts cannot seem to work in a way that gives structure and force. What does?

The goal of support for aviation emissions-reduction policy in developed countries should (presumably) be the ability to put sustainable fuel into aircraft tanks. Full stop. If only it were that simple. Keeping in mind that all support policy seems likely to continue only at the level of national (and perhaps even more local) government, what are the rational political and economic aims that we should consider achievable? Let's clean the blackboard (whiteboard today, I suppose) and start over with working up a policy approach. These policy notes will reflect more specifically some of the policy attributes that we outlined in Chapter 11.

To begin with, we must recognize that the capacity for funding public research, development, and other ways of supporting new fuel initiatives is not equal among states. Developing new fuel options exploits wealth and technical knowledge, resources that are found primarily in developed or rapidly developing countries; such work is being done in the US, EU, China, India, Brazil, Canada, Australia, Japan, New Zealand, and a few other countries, at various scales.

Efforts to support new fuel development must appear as a line item in someone's budget. So the question becomes one of establishing which domestic policies can be applied in countries like the aforementioned that will allow goals for *everyone's* commercial operations to be met. Thus it is necessary to discuss the options for making the fuel universally available, both inside and outside the borders of the countries most intensively supporting alternative fuel development. The preliminary work being undertaken by a handful of countries (and for which we sincerely thank them) must progress in a way that achieves commercial scale results for them, and somehow also carries the benefits into other parts of the world. So we must look first at the kinds of fuel development support policies that individual countries can apply. In terms of our ultimate aims, this is infinitely more important and productive than punishing airlines for *not* having sustainable fuel, or providing a mechanism for purchasing offsets as a substitute for burning sustainable fuel.

Supporting Technologies That Might Work

Basic Research

In most countries public support for alternative fuel development is most feasibly tied, in political terms, to the question of security of energy supply (or even self-sufficiency) and stability of price. What fuel support policies could be useful in this kind of scenario?

First, without question, the ongoing commitment of public money to unfettered 'blue sky' research is absolutely essential. It seems obvious that a broad spectrum of support programs for science that explores energy ideas at the very vanguard is key, and must always include support in the form of funding for basic, non-targeted scientific research, which ultimately provides the body of knowledge used in the development of new technologies. Governments all over the world recognize this, but some recognize it more than others. We must hope and assume that research establishments will always receive needed support.

As an aside, let us remember that knowledge production is vulnerable in a market economy; the economic value of a concept is unknown until it appears and is acted upon in the market, so it is hard to set a budget for the creation of more abstract or dawning ideas. Shortsighted governments may focus only on research that has particular, immediate technology application, thus reducing the possibility of discovering new and even better ideas.

Targeted Support (and its Traps)

But of course, we do also need support for specific fuel technology initiatives. These have historically been funded with public money in a number of ways, including the subsidization of technical development that targets work at various stages: lab work, pilot projects, commercial demonstration plants, and the implementation and even the construction of commercial-scale production facilities.[1]

It is essential that support for real solutions continue, but we have to ask if we are doing it in the most productive way. Supporting specific initiatives requires recruiting and vetting the various candidates out there. What purpose is served in having government do that kind of hunting? Even before a technology receives a penny, public resources are spent in searching for it. The results of the application and granting process can also be somewhat ambiguous and arbitrary; good ideas may get left by the wayside due to ineffective presentations, political pressure, lobbying, or other competitive imbalances. Partly, this can be attributed to matters of political constituency, and also to the fact that the success of appeals for support

---

1 The US Department of Energy/Division of Energy Efficiency and Renewable Energies *Financial Opportunities* website http://www1.eere.energy.gov/financing/current_ opportunities.html outlines the kinds of initiatives for which department funding may be available.

depends upon who is making the application, and what resources they can bring to bear in promoting their idea and lobbying effectively. The governmental funding agency must then strive to get a clear picture through an informational thicket that results from the inconsistency in various fuel developers' ability to describe and promote their ideas clearly and effectively.

Even when policy people try to be very careful, they still end up in the business of selection. In the case of biofuels and other alternative fuels, choosing a fuel technology is inherently difficult and can be made even more so by other, antecedent decisions—selecting a biomass for a biofuel, for example. If internal, national, and regional considerations restrict the support choices to a particular biomass or biomass type, or to a specific technology, other worthy options that apply different materials or processes will be pre-empted. And if choosing specific biomasses or technologies becomes too hard, do we consider supporting every promising combination of biomass and technology? Can we afford to do that? Probably not. The degree to which any option should be supported depends on somewhat subjective assessments of too many factors and too many candidates with potential. Making even general decisions puts governments in the position of perhaps pushing particular solutions, and eliminating others.

Policies evolve, of course. The US is a good example. The biofuels-promoting *US Renewable Fuel Standard* (*RFS*)[2] was first established with the enactment of the Energy Policy Act (EPAct)[3] in 2005, and was expanded into *RFS2* with the 2007 Energy Independence and Security Act (EISA)[4] (Schnepf and Yacobucci 2013). While carbon and greenhouse gases (GHG) were certainly a concern when drafting of this legislation started, it is fair to say that the bigger emphasis was domestic energy self-sufficiency, stability of fuel prices, and rural development. Whatever concerns led to this sort of policy, the result is instructive. This is an interesting case of a government trying to write good law in a very difficult area, and its struggle to achieve balance and compromise by establishing specific and complementary goals is apparent. But the politics of the regional vested interests are also evident, and the legislation inevitably becomes labyrinthine in trying to satisfy particular stakeholders. The kind of energy policy that the US has pursued cannot be written in neglect of matters of regional economic development, farm voters, and 'big agriculture' interests. As a result, the legislative process talks itself into distinct areas of technology and feedstock that are promising, but that are also supported in enough regions and segments of the economy. It then commits to them—often with insufficient further thought about their real overall commercial, environmental, and social viability.

What we are discussing here is a popular 'push' model of government support for specific technologies and producers that is, in many respects, counterproductive. To gain support for fuel development expenditures, policy drafters must address

---

2   See http://www.epa.gov/otaq/fuels/renewablefuels/
3   See http://www.gpo.gov/fdsys/pkg/PLAW-109publ58/pdf/PLAW-109publ58.pdf
4   See http://www.gpo.gov/fdsys/pkg/BILLS-110hr6enr/pdf/BILLS-110hr6enr.pdf

the concerns of legislators who want to see a solution that boosts the immediate fortunes of their particular region and voters. But when policy caves in to these voices, it establishes an even more entrenched, powerful, commercial and political constituency. Better energy solutions will not be supported if they might threaten a region's now-embedded technology or biomass source, and so in turn threaten its whole economy, dependent on the important activity of raw material production, processing, and fuel manufacture. So when these decisions are made for political reasons, a small vested interest can expand and attract broader popular support in a region, making it difficult for policy to change direction as needed.

Supporting Technologies That *Will* Work

One way of using public money more efficiently might be to move to a 'pull' model of fuel technology support. This means putting policy in place that is less attuned to narrow interests and is designed to be regionally, technologically, and sectorally unbiased. Here, we would focus on goals of very high and very objectively applied standards of sustainability and economic benefit, in the largest context.

This is really the crux of optimizing a national sustainable fuel endeavor: How do we most constructively draw on the available and constantly evolving pool of fuel technology ideas? In this regard, it is a true benefit (albeit, politically challenging) to be able to stay neutral on region, technology, and feedstock: to just see the ideas without the other baggage. Such an approach situates policy aims at the appropriate level: it is left to more local jurisdictions to find and support initiatives that fit the larger bill and are of more value locally. Thinking about policies at national scales, how do we develop those ideas in a way that requires less government effort and provides better, more financially realistic results using the same government money? Literally billions of dollars have been spent in the US in support of specific fuel ideas. Are mandates and quotas achieving ultimate policy goals? No. At least the predominant view is that they are not. Could that money be spent in a more productive and objective way? Or, perhaps a more politically practical question—one that does not imply immediately offending all of those who have become dependent on current programs: Is there a better way to spend any new or additional money?

For the purpose of generating more discussion (and that purpose only), I am going to venture into the territory of actually making suggestions. In contrast to the current policy model, the simplest and most efficient way of providing support might be to simply tender contracts to buy fuel. Rather than the granting agencies chasing fuel development ideas and trying to sort out what is currently and potentially promising, perhaps the fuels should come to the funding: fuel technologies seeking support should, while meeting set standards of sustainability on environmental and social fronts, quantify their level of carbon reduction against a fossil fuel equivalent and bid for a contract. The last piece of the puzzle would be price and volume—large volume. This is essentially what we could call 'bid for offtake'.

The Farm to Fleet program recently inaugurated in the US is one example of this *kind* of thinking: the US Navy has outlined its desired total volume of biofuel, and alternative fuel suppliers will be invited to submit their bids (United States Department of Agriculture 2013). This is an interesting initiative, but it does not fully exploit the enormous size and geographical scope of commercial consumption. And potential suppliers must demonstrate that they *are* prepared, not that they *would be* prepared to supply product. Was public money used to help particular potential suppliers achieve other milestones? The suite of measures taken together may effectively screen out small, new, or start-up enterprises.

Let us understand better how a 'bid for offtake' system would distinguish itself from other kinds of support in terms of incentivizing the evolution toward a hypothetically perfect and commercially viable fuel. The power in such a system resides in a few important facts:

1. Jet fuel is fungible and can be bought by anyone, anywhere, and swapped for fuel that is locally available to them and meets their needs. Any carrier that wanted sustainable fuel for its operations could buy that fuel wherever it was sold.
2. The concentrated demand for jet fuel at many individual airports, or groups of airports that are near each other, can be great enough to use the capacity of dedicated local production facilities.
3. *Total* system demand for anything, including sustainable fuel, is (obviously) greater than local demand.
4. National-scale financial support currently deployed in other ways is capable of buying at what would be massive scale for individual producers or potential producers and potential investors.

Taken together, these four facts ensure that a sustainable fuel production facility's output would be both consumed, and paid for. They also mean that if policy offered potential producers the opportunity to bid for the right to have their output bought, they could base their bid on the likelihood of being able to deliver that product at a location convenient to their production facility. Unlike other pedestrian offtake agreements, the scale of the commitment could be huge and long. Support takes a different form: if even the best bids can only offer the sustainable fuel at a price above that of the petroleum fuel that is also in the system, government support means paying the higher price and putting the sustainable fuel into the market at the prevailing price (or perhaps at a small premium.)

The thesis is that this is a more useful kind of support program because all potential producers, regardless of size, could compete. They could attract necessary investment on the basis of winning the bid. They could establish major facilities that would be efficient on the basis of the location, agreed volume, and duration of the agreement. And investors would only come in if they perceived the probability that the project would be viable in the longer term. It would also allow for consortia of prospective producers to mount a joint campaign on the basis of blending their different respective potential outputs. This might be a much more fruitful use of public money in that it incentivizes the best ideas to come to large scale quickly. And if an idea fails, the flow of public money stops.

By extension, in such a policy scenario, another important criterion would be that a fuel technology was demonstrably amenable to carbon reductions beyond those promised in the initial bidding round. There would be no point in an investor supporting a technology that was certain to eventually fail to meet other producers' better performance on carbon reduction and price point in subsequent years. An ongoing program would be designed to ensure the efficient use of public money in creating *an evolution toward a more sustainable national fuel supply that became progressively lower in carbon and was able to compete with fossil fuels on price.* Remember also that bidders in such a scenario would likely be commercial fuel producers and blenders who would undertake the dogsbody work of searching out viable fuel technologies that could provide them with the best possible blend components, thereby perhaps putting smaller potential producers in play.

This general idea is not entirely new. In fact the EPAct of 2005 contained such a provision, specifying incentives for the production of cellulosic biofuels (United States Congress (109th) 2005). Section 942 (p. 119, Stat. 878–80) mentions 'reverse auctions' that would function somewhat like the fuel buying mechanism described above. The Act is more concerned with energy efficiency and price than emissions, but the mechanism is similar.

Would such an idea work? It would be hard for most people—including me— to know whether it has any practical worth, but the value of any suggestion is not necessarily in the aptness of its first iteration or even the general idea itself. What is sought in offering it is that it might aid in discussions that leads to something that *does* work. Because what we have now does not really work that well.

For example, most of the EPAct and EISA focus more conventionally on mandates. Criticism of this legislation surfaced in 2011 in the form of a report written by an appointed committee of the National Research Council at the behest of the US Congress. The report casts doubt on the ability of current policy to generate production and demand sufficient to meet the RFS2 goals. An important legislative lapse, in the committee's view, relates to ambiguities generated by the way that RFS2 is structured. The report describes many uncertainties that affect the fuel industry but one is particularly interesting. RFS2 establishes mandated levels of use of certain fuels (conventional ethanol, advanced biofuels, and cellulosic fuels) but these target mandates are actually modified quite drastically to conform with forecast production capacity, in order to prevent dramatic increases in the cost of fuels to consumers (National Research Council (US) 2011). There are other limitations of the legislation but one important end effect of this particular weakness is that there is no real assurance that there will be a suitable level of demand for these fuels at a price that makes it worth an investor's while. In other words the RFS2 mandates cannot, in the end, be achieved because the economic and political price would be too high.

We have been using US legislation as an example of alternative fuel policy. But every other national or regional jurisdiction is bedeviled by the same problems. The EU RED (Renewable Energy Directive) and ETS (Emissions Trading System) programs—ignoring the whole controversy about whether international aviation should be included—are other examples of the same kinds of policy deficiencies. And, much more importantly, we are using the US example, because the US is one of the few countries that have tried very hard to promote alternative fuels. Publicly funded support for this has been enormous in the US, and regardless of the goals that spurred the initiative, it has paid dividends in revealing possible advantages that would interest those who seek more sustainable energy sources.

Shifting to a point that was described as critical earlier, I want to say again that we are necessarily talking about policies enacted by those who have the financial wherewithal to fund them—national or local governments with access to a sufficient tax base. In general, we must rely on developed or rapidly developing countries with very large economies for the kind of support that is essential.

So, the four major things of which we can remind ourselves as we continue are that:

1. Sanctions, including market-based measures (MBMs), are problematic, divisive and, while useful in the context of state emissions reductions, are least feasible in the international arena,
2. Their application in the absence of ready means of reducing emissions negatively affects growth and therefore development,
3. Only developed and rapidly developing countries will be well positioned to positively incentivize and assist the development of sustainable fuel technologies, but nevertheless
4. The new fuels *must* become available to everyone, in *all* countries—eventually.

The Air Industry Can Take Measures on Its Own

Beyond the discussion about the limitations of current or planned policy or what policy could become at whatever scale, we must consider the avenues open to the parties most immediately involved in the pursuit of sustainable alternative sources of fuel. There has been progress there. But if the efforts of pathfinder fuel companies and client airlines are admirable, and even if they bear good fruit, we have to acknowledge that speed is an essential element of our larger success and it just isn't there. What can the airline and fuels industries do, themselves, to lever the policy-making capacities where they exist, get the fuel, and make sure that its availability migrates to all carriers everywhere?

The aviation industry is probably best served if it assumes the most active role that it can. Air industry groups, for example, have pleaded with states at their ICAO gatherings to move the agenda on MBM, because the airlines need those MBMs in the medium term to meet their goal. We have seen what has happened. Even this partial, interim measure to account for emissions is in limbo. The members of ICAO have understood for a long time that consensus expects them to take responsibility for this matter, and they endorsed the application of MBMs on emissions (International Civil Aviation Organization 2013b). But agreement has not been easy to accomplish. The debate about Common but Differentiated Responsibilities (CBDR) or Special Circumstances and Respective Capabilities (SCRC) is ongoing, and developing nations feel that the activities of their international carriers, as part of their national assets, should not be subject to the application of an ICAO-established MBM regime. We have discussed the facts related to this whole matter of the Chicago principle of equal treatment being in conflict with the inapposite recourse to CBDR when it is sought in an inappropriate form. I mention that again now because it demonstrates very specifically how the needs of the industry itself are not being served particularly well, and points out how important it is for the players to consider what direct control they have over their own destinies.

The point is not that one faction of states or the other within the ICAO community is wrong. This is a fundamental disagreement on points that have merit. It is reasonable to sympathize with everyone's aspirations and points of view, but there is no escaping the logic that reveals the problem: We cannot confidently relegate the matter to an international body specifically tasked with controlling international air issues and ask it to consider each country's international air travel emissions individually. Probably compromises will be crafted and agreements will be achieved, but not necessarily—and not quickly. And if it turns out that states cannot dictate emissions reductions to others without creating legal confrontation, or the threat of pan-sectoral trade conflict, or even low-grade bilateral tensions, who can?

I would say that many of the people engaged upon the challenge of framing useful international aviation emissions policy, including delegates to ICAO, have fallen victim to the same outlook that preoccupies the drunk who has lost his keys (old joke): An inebriate stumbles around under a street lamp peering at the pavement, looking for the keys to his house. He is aware that he dropped the keys over by his front door, but prefers to look for them under the street lamp because the light is better there. When it comes to ideas for compelling emissions reductions, we know much more about national or regional or bilateral cap and trade schemes than anything else, but these may not be the best ways to tackle our particular problem; no matter how hard we look for an answer in that quarter, it is not producing solutions—and the struggle is paralyzing us. Maybe the answer should not be cap and trade, or even offset purchase as it is currently envisioned, or any program that is applied solely by governments—individually or collectively. Though every policy idea must involve governments, perhaps the answer is something different and nearer our front door. And while we are at it, perhaps the focus can be more on the long-term essential need to eliminate emissions and less on the interim, albeit desirable, goal of *compensating* for some of those emissions. Where will help lie if the industry cannot help itself?

A Self-imposed Obligation

Let us clean the whiteboard again and ask what the industry could do about emissions reductions that would satisfy the following outstanding matters:

- Capitalize on national research and development (R&D) efforts in a way that was fair, equitable, and supported by the countries doing the work,
- Extract some levy from the international aviation sector as an incentive for action, but, simultaneously,
- Recognize the advantage and onus of relatively wealthy states,
- Allocate benefit from fuels development to least-wealthy countries on the basis of ordered priority, and
- Work with a better national policy model; perhaps 'pull' rather than 'push' for supporting fuel initiatives.

Suppose there *were* a shift to a pull model of R&D support at national levels, and that its goals more specifically targeted the production of certifiably sustainable aviation fuels, along with other desirable products. In any event, pull model or no, given the number of new, sustainable fuel initiatives nearing commercial viability, we can reasonably hope that national or regional policy approaches to supporting new fuel technologies will produce results.

Now assume a different 'cost' or 'stick' approach—not cap and trade or offset/credit purchase. Assume rather that enough members of the international air carrier community agreed to a mechanism that allowed a small charge to be applied on a unit basis against any conventional petroleum fuel burned in international operations, regardless of who burned the fuel (profitable or unprofitable airline, based in a wealthy or poor country). We know that this would not be popular with the governments of non-Annex 1 countries (or many other countries for that matter) but it is an idea where the sector effectively taxes itself. So, to satisfy one important element of controversy, nothing here would violate Chicago Convention principles; all airlines would be treated equally—and we would have a stick. To address the other problematic element (CBDR), suppose that revenue from this scheme were used to bring nascent fuel technologies to the countries that were most in need of such fuel infrastructure—from all airlines to those countries with the lowest level of wealth on the basis of Gross Domestic Product on Purchasing Power Parity (GDP-PPP). Tying the bow, this directs the revenue toward the problem and not into a government's pockets; the money would go to the fuel technology developers who were able to make their systems work in those countries. The least wealthy countries would get the new fuel technology benefit for free. Note also that the economies of the (developed or rapidly developing) countries that had subsidized the generation of these technologies would welcome their expanded customer base; money they had spent to solve their own fuel problems would be recovered by having their companies selling those solutions to the IATA scheme.

How much money could a fee raise? The international aviation fuel bill for 2014 is estimated at $212 billion, and represents about 30 percent of total aviation operating expense (International Air Transport Association 2014). This (76 billion gallons) is currently almost all conventional fossil fuel; let us assign a benchmark carbon value of 'carbon 100' to it, and establish a charge that relates to carbon content. A baseline one-tenth of 1 percent fuel charge on this 'carbon 100' fuel would generate approximately $212 million; passing this fuel charge on to the customer would add about $0.32 per $1,000 to the passenger or cargo client's bill (based on current air transport use statistics.)

This could be a progressive and ongoing program. The revenue stream could be directed to the countries that needed the most development assistance until their technology and fuel production capacity met their uplift requirements, and then the support could be shifted to other countries. As fuel in various countries moved to lower carbon levels, the rate of the fuel charge would drop, though perhaps the total amount of monies collected would rise due to the consumption demands of increasing traffic. The rate of charge could be adjusted over time to ensure that revenue would continue to flow from this industry 'tax' as long as the need to fund new projects existed, or until the opportunity to implement further carbon reductions neared its limit. Then it could be tuned to a maintenance level, according to the availability of carbon-reduction solutions and the nearness to the goal of effectively zero-carbon fuel throughout the world. Since solutions representing good carbon-reduction capacity are already technologically viable, the initial and long-term rate of charge may never need to be more than slight.

Every single country would derive benefit from such a scheme. This is far better than any carbon charge arrangement that sends the money generated into national general revenues. This is taxing uniformly and to a solution, distributing benefit *as needed and according to principles set out in the United Nations Framework Convention on Climate Change (UNFCCC)*, to which most countries are signatories. It satisfies Chicago provisions by treating airlines the same while entertaining CBDR needs by treating countries differently. It would cost airlines money, but *any* scheme would cost airlines money—and airlines want a scheme. This would also mean a ready market for fuel technologies in which Annex 1 countries have invested, and a more rapid payback on subsidies that might have helped in their development. If international carriers voted to impose such a scheme, a formal protocol recognizing the agreement should even pass in the ICAO forum.

Any such initiative would fit easily with the kind of national pull policy fuel technology bidding ideas described earlier. Such a plan would return to the economy of a developing country more than the value of the tax imposed upon the operation of its international air carriers. But even wealthy countries would see any gaps in their sustainable fuel capacity filled by this arrangement, because they would be encouraged by the growing availability of new fuel technologies being developed within their borders—*if* they ensure that domestic policy continues to foster the development of such new technologies.

Perspective

Again, suggestions are for the purpose of generating dialogue. All of the material so far in this chapter is to the point of bringing one perspective and some ideas to the discussion with the goal of helping to assure that there will *be* a discussion; so much seems to have ground to a halt. Regardless of what comes of policy discussions, we should remember a few things. The first, already mentioned, is that the industry's needs and wants should be respected. The second is that the industry itself should seek control of its own destiny more actively. The third is that we should look everywhere for ideas. Lots of policy work in other areas may hold ideas. Lots of people have examined policy questions objectively and written volumes of material.

> I spoke with Paul Steele, who is Executive Director of the Air Transport Action Group (ATAG), and also Senior Vice President of Member and External Relations, and Corporate Secretary of IATA. The conversation was extremely helpful. We did not discuss the foregoing ideas about an international airline 'tax' because it was worked out after the time of our talk. In any event, reaction to proposals such as these must come from IATA's whole constituency. This book focuses on fuel but of course in international aviation, the most popular topic of discussion is the turmoil at ICAO *vis à vis* market-based measures. We talked about both. But in all respects, and regarding the whole area of the stalled process of international policy development, the positions of these organizations are crystal clear and easily accessed.

There are two general considerations that should be emphasized here. If it is true that the air industry and its potential fuel-manufacturing partners must be listened to very carefully at this point, it is a critically important fact that the air industry itself maintains a unified position that aviation sustainability is an essential goal. This willingness to be public in commitment and action is, in every way, more important than the difficulties that have cropped up in trying to agree international policies on the matter at hand. The second consideration is the flip side of that same coin: If the industry position is to be taken seriously and remain credible, the industry itself must act to give it effect. There are certainly international bodies, like ICAO, that might have offered quicker and more concrete guidance, but industry commitment exists separate from the help that it might have received in achieving it.

Our discussion did cover the deliberations at ICAO, of course. Steele pointed to an industry proposal presented as a Working Paper to ICAO's 38th Assembly, 2013 (International Civil Aviation Organization Executive Committee 2013). It suggests that it could be feasible, and preferable, to treat individual carriers differently on the basis of rates of growth rather than location of base of operations. Regarding this suggestion, there will be questions that can only be answered through discussions. One relates to the nature of airline traffic growth: some comes through attracting new travelers, some comes through attracting competitors' customers. It might seem odd to offer operating cost relief to carriers that are enjoying a cost advantage that helps them attract traffic from an established carrier. One also wonders if governments will start plowing capital into their national carriers in order to manipulate the growth criterion that would bring the prospect of more commerce in and through their hubs. In other words, while growth is normally transient, growth-based relief from carbon accounting could create some instability in that dynamic. So it is hard to know if the idea would gain traction. Neither this nor any other idea on MBM swayed enough minds to win immediate approval at the 38th ICAO Assembly of course. Nevertheless, just the fact that the ATAG constituency—and that includes IATA—considers this a useful approach is important. We do know that the application of MBMs is essential to carbon-neutral growth for at least an interim period from 2020, and we hope for acceptance of an idea before then.

However, no matter what happens at ICAO, the pursuit of better fuel simply must continue (perhaps even more urgently unless international talks get going properly) and is affected by principles like CBDR/SCRC in a very different way. The positions taken by both ATAG and IATA have been made clear and they are very focused on moving toward sustainable fuel, as we will see in more detail in Chapter 14.

Other Voices

I spoke with several people who, while interested in certain things that have been discussed in this book, have little to do with aviation directly. This somewhat freshened my perspective, and I would like to mention one conversation in particular. Chris Turner, as a journalist and author, has researched extensively in an effort to figure out what technologies seem to help to make human enterprise less disruptive from the point of view of sustainability, and what policies best support the technologies. While he has not examined the commercial aviation challenge in any special depth, he does have great insights regarding energy more generally. When I asked him for his impression of the general policy approaches that produced results most reliably, he made a number of points. And I think that they are relevant here.

First, among the assortment of 'sticks' that can be wielded against the carbon-emissions monster, his view is that straight charges (acknowledging that tax is a dirty word in many quarters) are by far the most effective. Cap and trade has proven less effective as Turner judges the EU ETS, for example. Too many variables affect markets, including carbon markets, so price signals can be inconsistent, inappropriate, and ill timed.

But Turner is not entirely enamoured of the stick as the most effective tool anyway, and this brings us to his second major observation: Of those things that governments can do to encourage the development of energy capacity, purchase commitments have proved the single most effective. In electrical grid power, the feed-in tariff (FIT), which commits for the purchase of more sustainable power at a premium, is the policy tool with the best track record. Surely nothing shows incentive policy success more than installed capacity, and one point in a recent report prepared for the US Department of Energy (DOE) points out that installed capacity (in solar-photovoltaic for example) is high in Germany, where FIT is most aggressively applied (Barbose et al. 2013). This argues that while assistance to both technology developers (in the forms of basic research, or start-up grants), and also to consumers (in the form of installation subsidies) are all indeed important, a piece of paper that commits a power provider or government agency to actually *buy* the power that is produced is the element that tips the scale.

We have learned that policy cannot be simplistic, and the earlier Spanish example in Chapter 11 helps us understand how easy it is to make a mistake. But I think that Turner's larger perspective is valid: If we have a goal, we should go easy on the stick and build capacity.

Summing Up

An important failure of all current emissions limitations policies (whether in place, proposed, or merely mooted) is that they do not promote valuation of carbon in a way that provides the highest appropriate incentive for the greatest feasible reduction. But it could be argued that such valuation is effectively impossible: We cannot know exactly the cost of reducing the negative externalities, and we do not know the pace at which we can remove value from an economy, a company, or a sector without causing it to fail, thereby diminishing wealth generation faster than we can develop alternative wealth creation activities to inject wealth *into* the economy. We do not know enough about what technology innovators can accomplish, nor do we know what their novel products will cost—especially in the context of an evolving market for those products. This limits our ability to both apply sanction and assist technology development intelligently and appropriately.

All of the preceding argues that the imposition of policy cannot rationally reflect the reality of climate threat along with anyone's ability to change, based on a given day's snapshot of all the economic and environmental variables. And so it follows that a sudden, wholesale change to the important economic parameters of an industry or sector or whole economy cannot be entirely appropriate, practical, or desirable.

And yet, we must get on with the job, and the only way to do that is to focus on pace: If policy should start incrementally, yield data that allow us to measure achievement, and then be recalibrated, maybe we are not as lost as we might fear. As long as policy produces no instabilities, it can be gradually ramped up as the economy's ability to eliminate externalities increases. It can also be inferred that where penalties are necessary, if cost associated with carbon emission is introduced gradually, it will, at some point, become more directly related to the costs of avoidance.

Stick measures, however, can perhaps be a risky gambit for a couple of possible reasons. If there are no imminent fuel technology solutions, there is no way of avoiding carbon costs. Second, if an industry's ability to pay is so compromised that imposition of a meaningful cost creates important liquidity effects and provokes an evolution toward industry winners and losers, then such measures cannot be considered fair or acceptable. Fortunately, the air industry does have some alternative, sustainable flight energy options available to it, but the question of how to pay for emissions still remains.

One additional point to remember is that the air industry is under chronic financial stress trying to improve efficiency. Procurement of new aircraft with new engines, where development costs are great, constitutes never-ending pressure on balance sheets. The broad application of substantial carbon restrictions through the imposition of restrictive caps (with the option of trading for emissions permits) will be almost impossible to achieve in an equitable way. If the caps are restrictive enough to impose serious carbon limitations, they will be entirely disruptive, hinder the industry's capacity to provide the wealth that funds improvements in carbon reductions, and create winners and losers among states and airlines.

From the point of view of policy-makers trying to decide on mechanisms, this makes for a difficult challenge. Is it possible to see airlines charged for their emissions and then have that money spent on solutions? How? How much? If we do not know how much an action will cost, or how it should be charged, or how to make sure that all players are charged equally, and we know that, even collectively, airlines cannot pay much anyway without a 'winner and loser' kind of collapse, the whole initiative seems impossible. Thus far, work on imposed policy approaches has encountered barriers represented in the discussion above. Of course it is still an option to let the aviation industry realize that change is inevitable and allow individual airlines to go out and find their own supplies of new fuel, but those efforts would probably be inadequate, and they would favor the most profitable companies from wealthy countries.

As a consequence, efforts in development and procurement of actual low-carbon fuel have been limited in magnitude, slow to progress, and often carried out privately by those carriers that can afford the cost of the public relations benefit of being able to say that a bit of their flying is conducted using alternative fuels; in most cases, those alternative fuels depend on technologies that enjoyed public support in their development. That last point is important to remember as we advance this discussion.

I anticipate that we will probably always end up with both support and punishment policy mechanisms. But if we need both, it is absolutely essential that their application be carefully coordinated in a rational pursuit of policy objectives rather than being foisted upon the economy in a way that sees the carrot and stick operating inconsistently or even at cross purposes. At this writing, there seems to be some support for the general idea that MBMs applied to international aviation should be in the form of offset purchases. These would be used to meet the (interim) gap between actual carbon-reduction ability, and growth in the sector. One avenue to consider would be to create offsets that represent reduction in aviation emissions where they would otherwise not be required. That would include domestic air operations in non-Annex 1 countries and (should the ICAO Assembly agree to route-specific exemptions) international air operations into such countries.

The Long and Winding Road

National governments, the airlines based in those countries, the organizations that represent those airlines, and the ICAO Council (constituted of national delegates) have to come to terms with some of the facts as we all now understand them. Regardless of what is decided at the ICAO Council in the next few years, the most important challenge is the development of policies that will serve over decades in moving aviation to 100 percent sustainable fuel. If MBMs, the dominant issue of the day, have not dwindled to just a sideshow in the longer term, we will be in a lot of trouble.

There has to be acceptance of the need to provide incentive for the introduction of alternative, sustainable sources of flight energy. The way to do that does not have to be any of the ideas that have been presented earlier here, but it has to be new, and it has to be big. It has to give every country, rich or poor, a reason to want to participate. It is time to give imaginations license. Remember, the challenge is not merely to reduce emissions *somewhat* and in a *somewhat* sustainable way. The challenge is to achieve, within a few decades, an essentially perfectly sustainable fuel that contains virtually zero carbon and also respects other environmental and social concerns. We do not have the luxury of dismissing this goal as either laughably impractical or optional; in real, hard terms, it is absolutely essential.

Chapter 14

Discussion of Current Attitudes and Efforts

The challenge is mighty. Policy is supposed to help. What is it helping? Is there really fuel that is sustainable now? Who is buying it? Why? What are industry groups saying? What do the experts think of our chances? If we take everything that we have considered so far and attempt to paint a picture of aviation's sustainable flight energy project, what does it yield in terms of line, shape, color, and texture? There is a financial, commercial, economic, political and policy context for everything that might arise. How is it all fitting together? Yes, nascent commercial enterprises are delivering, or are on the verge of delivering, more sustainable fuel product at usable volumes, but will the capacity build? In all of the ways that we have considered the subject, what happens now?

It will help us if we examine organizations and personalities that are poised to make notable differences. If we assume that change is happening but not yet at an appropriate pace, we would want to know what has to occur in order for that pace to develop. Policy does matter, and if we keep policy questions in mind as we canvass the views of key actors, we can perhaps see how they are assisted or resisted by policy at various scales and how certain initiatives, or changes to the policy landscape, might alter the larger situation as well as the chances for commercial breakthrough and viability. We will start this chapter by examining the world in which the experts live. Then, we will see what they have to say.

Fueling Progress

The economic factors that one might associate with the whole matter of bringing in an alternative supply of aviation fuel are not easy to list and summarize. First, flight energy economics have to be understood in the context of all the various economic factors that affect everything; the economy is much like the environment in that respect. Acknowledging that a completely integrative analysis is beyond the scope of this book, it is still important to consider a few things about the global economic context for the discussions on the specifics of the air travel industry and also the energy industry.

Convergence

One issue is the matter of economic convergence. It is of interest because we know that development goals affect policy discussion; that aviation supports development; that development must be sustainable; and that fuel is fundamental to that mix of issues. Some economists would say that disparity in wealth between countries (or even *within* countries) affects the rate at which wealth builds: poor people and poor countries can get (relatively) wealthier faster. That would mean that disparity in wealth constitutes the essential driver for resolving that disparity: different groups of people with different levels of wealth tend to naturally move toward a common, higher level. The same would then be true for economic indices and factors that characterize wealth: the source and use of energy, and the efficiency and emissions associated with that energy are such factors, and they are key to our larger subject. Critically, so is air travel itself. In general, if societies are destined to be similarly wealthy, they are destined to have similar access to the things that drive that wealth.

There is some evidence that such an equalizing economic effect exists. We could hypothesize that people and societies that aspire to more wealth try harder, or have lower costs. But whatever the mechanism, there is a substantial amount of theory behind the idea that economic differences naturally get rid of themselves, that levels of GDP drift higher and come together. The problem is that they do not *always* converge—not necessarily. A very useful summary that displays this fact and offers insights into possible reasons for it is provided in a Lowy Institute paper (McKibbin and Stegman 2005). Countries that are similar in other respects *do* tend to converge in levels of wealth. Countries that are dissimilar do not. In this case 'other respects' seems to cover a wide array of factors that might relate to technology and the capacity to absorb and exploit new technology.

The paper uses the well-known IPAT formula (*Impact = Population × Affluence × Technology*) (Ehrlich and Holdren 1972) as starting point. The formula itself is arguably simplistic. But McKibbin and Stegman elaborate it into the form we have already discussed: *Emissions = Population × GDP per capita × emissions per GDP*. And regarding the last factor, we know that 'emissions per GDP' depends upon the amount of energy per unit GDP and the amount of emissions per unit energy.

A simplistic conception of convergence theory implies a certain evolution: that what we might call the development 'vectors' of societies vary. With different starting points, they are headed toward the same economic destination. That *is* simplistic, because what we read in the Lowy Institute research is that such a result would seem to require that developing societies either have (or can acquire) not only the technologies that have created advantage for the wealthy societies in the past, but also the technologies that will advance all societies in the future. Further, that requires yet another condition: that the people in a poorer society are as prepared and equipped to adapt and apply these technologies as those in wealthier countries.

Aviation has served as a development enabler and as an ongoing service to developed societies—our whole discussion assumes that this will continue to be the case. But now commercial flight must be that enabler for everyone; every necessary or advantageous level of commercial flight service must be available, together with all of the technologies that both the wealthy and rapidly developing worlds will create to make it environmentally practical. This will not happen as some sort of 'autopilot function' of global economic evolution. We cannot rely on an abstract conception of convergence that does not conform to reality. Aviation—and a perfect, sustainable form of aviation at that—will not, on its own, become something that everyone will enjoy. Data demonstrate very clearly that all things are not equal; never have been.[1]

I mention all of this now, when we are talking about the economic viability of a change in flight energy source, because the implications are clear: Either the undeveloped world stays undeveloped and we abandon, as fiction, the 'developing world' concept and any notion that reasonable standards of living can be brought to more and more of the global population, or we recognize that human and economic development must progress. There will be a truly enormous need and market for technologies that reduce emissions intensity of energy toward zero. But the capacities that will allow development to be environmentally and socially benign have to be brought to those who need those tools by the developed and rapidly developing worlds. Having development targets does not mean that the best technologies will be chosen to pursue them. Likewise, the mere existence of new technologies does not mean that they find their way to all those who need them.

Efficiency Versus More Fundamental Change

And, again, efficiency is not enough. That is an important economic fact that should inform policy. Huge amounts of capital are being poured into the effort to improve the efficiency of fuel use. Airlines support this because fuel, no matter where it comes from, must be paid for. Aircraft manufacturers enthuse because it affords an opportunity to sell new aircraft technology. So do engine manufacturers. And it does produce a favorable result—but not favorable enough. The aviation system should certainly be made to use fuel more efficiently, but policy formulation should also examine the deployment of capital by industry and government, and ensure that reducing the *emissions intensity of energy* is enjoying an even greater amount of attention. At the extreme, we could even say that the energy intensity of GDP does not really matter, if the energy is sourced sustainably.

Policy must comprehend technological, economic, and social context. If it is necessary to change technologies, the question of whether any society is prepared

---

1 The United States (US) Energy Information Administration's *International Energy Statistics* website http://www.eia.gov/cfapps/ipdbproject/IEDIndex3.cfm contains tremendous amounts of data on such things as the production, consumption, intensity, source, and cost of several types of energy in different countries over many years, along with statistics on emissions, energy markets, reserves, and a host of other variables. Data can be displayed, downloaded and organized in ways that provide useful comparisons of different countries' use of and exploitation of energy, and outlines of their development paths.

to implement the requisite new ones is entirely open. What are these technologies? How do they work? How many people do they need? At what cost? What materials? If we do not know everything about the new technologies that must be deployed, these questions must inform our analysis.

Among the suite of characteristics that we can envision for a growing, prospering society, controlling the carbon expended in flying a person around is very challenging. A country that wishes to participate in greater development, and those who would help that effort, must ask what technologies answer the carbon-reduction questions. And we must think not only in terms of the dictates of scientific realities, but also the social realities of (perhaps) low education, low mobility, cultural reliance on subsistence agriculture, a history of local conflict, and any of a host of other factors that might affect local exploitation of such technologies. So the questions become:

1. What menu of technologies can be developed and made available?
2. Which of them can be deployed in a given location in an environmentally and socially non-disruptive way?
3. What must accompany the technology to allow the local population to engage with it?

The stated need to be able to select from among options clearly implies that there will likely never be a unique, ideal, one-size-fits-all sustainable flight energy fix. So we must also hope that everyone understands how complicated it will become to replace fossil energy when there must be a solution to every energy problem in every location, and that each selected solution must address *all* of the implementation and sustainability factors comprehensively. Proper certification criteria can perhaps now be seen as the absolutely essential filter to be applied in selecting solutions.

The Buck ($) Starts Here

What technology? Where sited? Advantaging whom? Those questions bring us to the distinction between larger economic purpose (why we build capacities) and the practical commercial exigencies (who will make money by exploiting a need?) Economic viability is an essential part of sustainability, it has become very clear how important that fact is. It is true that economic practicality for the world or for a society is not the same thing as profitability for a certain company, but it is essential that the project of equipping the world with new sustainable aviation fuels must become commercially worthwhile.

Public support for commercial development of fuel may be necessary; we have already argued that it *is* necessary. And where a government has decided that new fuel initiatives are economically vital because the costs of *not* having such a fuel are too high, development of fuels or even their ongoing production might be subsidized. That's fine. But unless we see never-ending public support for aviation fuel—and I suspect that most people would not agree to that—we must pay a lot of attention not only to larger economic merit but also to how that merit can become rewarding and practical for autonomous, commercial enterprise.

Leaders of the air travel industry and those who are in a position to enable its transition to being a sustainable and benign contributor to the world's development and commerce must decide what to do, and a number of opinions are presented in the rest of the chapter. But before we embark on this examination of ideas, initiatives, attitudes, and approaches, we should consider some particular factors that inform the economic discussion.

The Business of Sustainable Aviation

When we set out to discuss aviation and fuel economics in a book of this scope, it is more in the vein of removing the cowlings from a jet engine to see how complicated it is rather than to actually explain how it works. However, it is a worthy subject.

First, we have learned that the airline business is capital intensive, and we know a little about what it takes in terms of financial resources in order to park a large aircraft at the gate, ready for flight. The return on such capital must be a carefully calculated part of any airline's business plan. The costs associated with exploitation of that capital asset have a dramatic effect at the margin. When we think in those terms, the fact that air travel is also energy intensive, with fuel costs making up a rising percentage of total costs, is extremely important.

I touched upon a number of points that relate to economic questions with John Heimlich, who serves as Vice President and Chief Economist for Airlines for America (A4A), which is the trade organization that represents United States' commercial airlines. He updates and confirms again for us that fuel has now risen to 30–40 percent of airline operating costs. But that rise is not entirely predictable. If energy costs were known, the industry could make more realistic plans for capacity and price. The value of capital assets is affected in many ways by fuel energy cost variability. Generally, anything that drives up cost makes it difficult to exploit already-expensive capacity; revenue—and that usually means fares—must rise and, presumably, traffic reacts to any such rise. And yet, for all of the same reasons concerning the capital intensiveness of the industry, its ability to hedge against such risk is quite constrained. So fuel price volatility is probably a greater threat to the industry than simply high fuel costs. And it is certain that the industry responds to fuel price in a complex way. For example, while elevated prices for fuel and the consequent need for higher revenues might certainly dampen demand for travel and erode orders for new aircraft, at the same time, aircraft that exploit newer technologies are more fuel-efficient and so there is some pressure to acquire such new aircraft.

How these countervailing factors interact and how that interaction affects manufacturers' order books depends on many things, including where a manufacturer is in the development cycle for new product, and how airlines view potential traffic prospects and their assumptions about such things as future fuel price and future cash flows. So we can be assured that fuel price volatility creates headaches for airline managers and that these headaches tend to dampen enthusiasm for industry development. It can be argued that the aircraft capital commitment cautions, coupled with unstable fuel prices, abet some of the stutter-start nature of aircraft development, purchase, and deployment. Industry discourse reflects this preoccupation. Coverage in aviation news periodicals shows this. An excellent example is a contribution by Alderman & Company (2011). They talk specifically about how aircraft orders respond to fuel price spikes. Manufacturers and their client airlines tend to swing each other around strenuously as a result of being coupled through such things as energy cost instability; they amplify its effects for each other. Illustrative of how rarely the economic stars align for air carriers in this complex dynamic, Heimlich points out that 2013 was the first year in over a decade in which US GDP grew while jet fuel prices diminished.[2] Expensive airplanes that were more efficient arrived at the same time that revenue grew, making the bets pay off. This goes deeper into the economy as well, affecting providers of services and infrastructure and also the actions of both current and potential airline customers (because the price that will be charged for future air travel is less knowable, and some clients make provisions for other ways of doing their business—teleconferencing and the like).

All of this is to say that, while our primary concern in this book is the sustainability opportunities offered by different sources of fuel, the economic and financial effects of its introduction, and particularly the opportunities offered for security of supply and (above all) stability of price, could be much stronger drivers. Some people in industry might be prone to framing the sustainability matter this way: 'Yes, it is desirable, but is it affordable?' At base, and in terms of the way the world is coming to regard sustainability, that should become a nonsense question: security of supply and stability of price imply the more important point, that we cannot afford to be *without* alternative fuel; and it may turn out that the direct economic advantages of having alternative sources of flight energy are what drive us most strongly in the direction of sustainability. And, over the long term, sustainable, alternative aviation fuel may be able to demonstrate to other sectors and to policy-makers that 'sustainable' really does mean 'possible' in pure business terms.

Commercial Viability

Nevertheless, new sources of fuel must fit into the larger economic picture. Part of the appraisal of more sustainable fuel production options will be to determine

---

2   Trends in aviation fuel prices can be viewed at http://www.transtats.bts.gov/fuel.asp?pn=0&display=data4 Trends in GDP can be viewed at http://data.worldbank.org/indicator/NY.GDP.MKTP.KD.ZG

how cost of production for a given fuel initiative plays against the industry's need and ability to pay. The economic dimension is meaningless in the absence of some understanding of the factors that influence the global market into which the fuel must be sold. Furthermore, economic factors will decide whether or not a flight energy project will be undertaken regardless of how well it might score on other criteria. So we really need to understand a little bit about the economic aspects of our larger subject.

Recall (again) two points relevant to economics and particularly relevant to policy:

1. Aviation makes a contribution to the global economy both directly and in support of other industries and activities. The Air Transportation Action Group (ATAG) claims that aviation is responsible for 7.5 percent of global GDP (Air Transport Action Group 2009).
2. The economics of converting to alternative, sustainable, fungible fuel from the point of view of distribution entirely favors aviation, where it enjoys access to a production and distribution chain that goes right to the aircraft tank.

The latter point implies that aviation is more easily targeted for conversion to sustainable liquid fuel than any other mode, and the former points out that it is more aptly targeted. We have already talked about that, of course. The reason that we make particular note of it here is so that we can think more competently about the barriers faced by those who are tasked with getting sustainable fuel into aircraft tanks. They operate in a world of competing interests.

A Note to Developed Countries: Aircraft Export

Countries with an aircraft-manufacturing sector experience an important amplification of these commercial effects. In the Aerospace Industries Association (AIA) *2013 Year End Review* it is pointed out that net exports of manufactured goods in the US, for example, are led by aerospace, with a favorable balance of trade for the sector of $73.5 billion in 2013. Civil aircraft, engines, and parts constitute about 88 percent of those exports (Aerospace Industries Association 2013). If selling aircraft is important, the question of whether the customer has fuel for them is vital. So fuel economics and sustainability affect not only US airlines but US manufacturers as well, in both foreign and domestic markets. From an aircraft export and balance of trade perspective, it is important for sustainable fuel to be available *everywhere*. Foreign customers for US commercial-aviation manufactured goods will benefit from local security of supply and stability of price for fuel. It is therefore to the advantage of the US economy to see global progress in new fuel deployment, not only in the US but also around the world.

Again, we speak often of the signal US case, but the same is true in other developed countries. A Canadian government document promoting Canadian investment points out that aerospace generates $41.2 billion 'across multiple supply chains in Canada' exporting 80 percent of its production, and that this activity is highly focused on civil aviation (Canada. Foreign Affairs and International Trade Canada 2013). It seems hard to argue that any of this is economically unimportant for anyone.

Just How Commercially Viable is it?

Having reviewed the economic value to the economies of whole countries and the global economy of alternative, sustainable jet fuel, what about commercial prospects? Are there fuel technology pathways and policy frameworks that present opportunities to secure sources of more sustainable flight energy in a way that addresses the fuel developers' and producers' needs for financial return, and also meets the air industry's requirement that the pricing of such fuels is stable? Can market forces move us toward progressively more sustainable fuel?

Considered in economic isolation, as sustainable, alternative supplies of fuel are developed, they increase the total supply of jet fuel, which exerts some downward pressure on price. But all factors are *not* isolated. Things will not remain constant, and we can assume that demand for the overall commodity will rise. Therefore, the relative market share of two sources of kerosene (fossil and sustainable alternative) will be determined by either or both of two factors: relative costs of production, and individuated demand. We can imagine some of the many conditions that might drive the former. The latter is sensitive to a number of things too: total demand (of course), public policy that specifies sustainable alternative fuels, and organization-specific corporate policy that settles on certain sustainability goals. Even when they understand their own part in the play, it is obviously hard for airlines, potential fuel providers, and those who represent and advocate for either, to gain any confidence about their comprehension of this dynamic, because of the complex interaction of the other factors. So figuring out the financial practicality of commitments to new fuels is tough.

And there are still more things that complicate the entry of a new fuel source. One problem is that since relatively little production capacity exists, contracts for alternative sources of fuel are generally fixed-price contracts on future delivery, rather than simple spot market agreements for supply: the fuel provider needs that financial certainty to manage its capital expenditures (for new processing facilities, for example) and its commitments to upstream suppliers who are in the same boat. In the case of biofuels, the biomass source commodities that a fuel manufacturer might use are also subject to market demand; the cost of land, energy, capital, and human resources that could otherwise be used in different ways are also market-driven, so the cost of biomass raw material to the fuel producer can fluctuate in the same way that petroleum-based jet fuel price would fluctuate anyway. This means that fuel producers must, in turn, have stable price agreements with biomass suppliers. This gets very dicey because until transition to a non-carbon economy becomes somewhat embedded, the biomass suppliers must purchase all of their inputs in a carbon economy that responds to spot price for petroleum. This commitment to the longer term at three different levels (user, producer, and materials supplier) means that the players will impose compounding premiums on their participation in order not to be caught behind the market at a future date.

We may as well remember here that this is one reason why the market will, and policy should, favor alternative fuels ideas that are less dependent on resources that lend themselves to fluctuations in value. And this gives practical merit to some of the 'food for fuel' arguments that are otherwise framed in a more ideological way. Whatever allows fuel to be produced without depending on land, valuable fresh water, and other resources that could be used for *anything* else, is preferable.

So, the first thing to understand about the mind of a fuel developer is that this one thought never leaves it: Risk costs. There must be accounting for risk. Generally, a premium must be calculated to cover for reasonable assessment of risk. An investor must balance that premium against staying out and thus eliminating a chance to profit. 'De-risking' alternative fuel enterprises is complex. Risk premiums are added at every stage of the process of getting fuel to tank. When all of the insurances required by all of the investors in all of the components of a deal are added up, they can constitute an enormous financial barrier. The 'natural selection' process of fuel production options will favor situations where:

- petroleum jet fuel is *relatively* hard to come by;
- the anticipated cost of a production facility is relatively well understood;
- the anticipated cost of the facility is relatively low;
- the fuel manufacturing facility can be sited relatively close to where the fuel will be taken up;
- the cost of feedstock processing is relatively well understood and stable;
- feedstock production is relatively close to a processing facility;
- feedstock processing is near to the manufacturing facility; and
- the cost of production of feedstock is known and stable.

All of that constitutes a unique model for each fuel initiative. Hardly any fuel will be produced in accord with such an idealized scenario, but coming as close as possible to such criteria will be a part of having a viable fuel enterprise.

And it does drive one kind of focus that is very active. Airports are almost always sited where people live, and anything that people or their activities produce as an *inevitable* and GHG-generating waste, that can be turned into fuel, is a prime target as potential feedstock. In such a case, the distances of links from feedstock production to processing to fuel manufacture are tiny. If the material is truly waste, the feedstock is cheap. So the big variables are the costs of processing and fuel manufacture. This is an enormous advantage in a business where feedstock production and transportation is generally the largest and most variable cost factor.

I would stress that the ideal waste material must be both inevitable and GHG-producing because if the waste, itself, is something that can be targeted for elimination or other use, the day will come when using it to make fuel will constitute an environmental and financial or economic cost. For example, if the waste from a certain industrial activity is truly valueless but is stable and non-degradable in a landfill, converting it to fuel would result in the ultimate release of carbon that would otherwise be safely sequestered. Alternatively, if it is economically useful in any other way, it can become expensive to use it for fuel.

One thinks of sewage and food waste. These things must be produced and must break down to produce CO_2. So it does not add any GHG to the atmosphere to convert them to fuel, use the energy, and *then* release the CO_2. Of course as soon as anything is commoditized, its value starts to be apparent. If sewage and food waste can readily be converted to energy, more than one enterprise will wish to use it. Then we move from 'How much will you pay me to take sewage off your hands?' (excuse the image) to 'How much do you want for your sewage?' (!) In fact, if it is practical to convert sewage and food waste to energy, it can be used to satisfy immediate, local community energy needs. On the other hand, if it *is* used to make fuel for a community combustion power plant, the flue gas—which constitutes a concentrated, exploitable source of carbon—can be captured and converted to jet fuel.

Remember that many waste materials can be made into fuel, but not all waste materials can be sustainably sourced or achieve real carbon reduction on a life-cycle analysis (LCA) basis. If carbon-containing waste, such as many plastics, is turned into fuel, that would constitute a net increase in atmospheric carbon. A good case can also be made for burying carbon-containing waste even if that waste comes from renewable sources like wood. It makes us ask if, for example, municipal solid waste (MSW) that is permanently buried in a stable anaerobic environment should not qualify for a carbon credit.

We will see that many of the most promising technology/feedstock combinations focus on waste.

Points of View

That is enough background. Now we should hear from those who actually make or use these fuels about which we have heard so much. If what follows seems like 'boosterism', recall that we are hearing from those who support the evolution toward use of sustainable aviation fuel. In some ways, it is quite amazing that there are any who do. There are certainly people who are interested in both industry and sustainability. But it is also surprising, in a way, that those who must run companies and make sure that they are profitable would have any appetite for facing any of this. But there sure *are* many individuals and organizations within the commercial aviation and fuel communities who seem to exude enthusiasm for the project of bringing air travel emissions to low levels in a way that is demonstrably, in other measures, sustainable.

I was impressed when I discovered the work being done by the Roundtable on Sustainable Biomaterials (RSB). I was even more impressed when I found certain airlines and air industry organizations committed to RSB standards. That fact is one that demonstrates how the air industry has committed itself, probably irrevocably, to change. In earlier chapters we presented what some will inevitably argue is an idealistically comprehensive view of how the concept of sustainability works. Later we find that the RSB standard actually embodies the precepts. Then we see that large parts of the industry want to see that particular standard embraced. The story of how the industry is engaged on the sustainability challenge offers that reality as the first important bit of the plot. There is acceptance of the idea that change is coming and change is good.

Every airline will eventually have to accept its obligations to the planet and society, but there are always leaders, and in this case they have often acted in concert so that their individual efforts are better supported. They form associations to present their ideas and goals to the public, government, and other organizations and industries. Or they join groups whose efforts are established on the right path. First we will examine some of the industry groups that are active and then we will meet some of the corporations and people who play in this sandbox.

Commercial Voices: Organizations

Sustainable Aviation Fuel Users Group (SAFUG)

The first of these that I will mention is the Sustainable Aviation Fuel Users Group (SAFUG), a collaborative organization formed among airlines that share sustainability goals. If you read the list of members below, you will conclude that that each one would have different reasons for signing on. Nevertheless, they are all working together and have adopted corporate policies that underwrite SAFUG principles.

Members:
- Airlines: Air China; AreoMexico; Air France; Air New Zealand; Alaska Airlines; ANA; AviancaTaca; British Airways; Cargolux; Cathay Pacific; Etihad Airways; GOL Linhas Aeréas Inteligentes; Gulf Air; JAL; KLM; Lufthansa; QANTAS; Qatar Airways; SAS Scandinavian Airline; South African Airways; Singapore Airlines; TAM ; TUI Travel; United Airlines; Virgin America; Virgin Atlantic; Virgin Australia.

Major, non-airline affiliates include:
- Aircraft manufacturers Airbus, Boeing, and Embraer,
- Fuel technology producer UOP.

The key plank in SAFUG's organizational policy platform is its member Pledge, whose introductory paragraphs note that flights operated by SAFUG's member airlines constitute about 32 percent of global commercial aviation fuel demand. Reading through the Pledge itself, we are made aware that the Natural Resources Defense Council (NRDC) and the RSB, together, were involved at the very inception of SAFUG, and that the Pledge involves commitment to broad sustainability guidelines which, in my view, essentially précis the RSB sustainability principles, and contain the phrase: 'These criteria should be consistent with, and complementary to emerging internationally-recognized standards such as those being developed by the Roundtable on Sustainable Biomaterials' (Sustainable Aviation Fuel Users Group 2014).

SAFUG, with its principles and the significant membership that backs them, constitutes a powerful commitment to really sustainable alternative fuel and common global certification criteria that would support that achievement. But older, broader industry groups also support just as strongly.

International Air Transport Association (IATA)

IATA is *the* well-established airline industry association. It represents some 240 airlines, constituting approximately 82 percent of the world's commercial air activity. It is involved in a number of interests and forums, so it is certainly not narrowly focused upon environmental and sustainability matters, but it does state its support for RSB and other principles of sustainability in securing alternative sources of fuel. In its *IATA Sustainable Alternative Aviation Fuels Strategy* document, it includes a qualifying phrase stating that such fuels would be 'for use in jet aircraft as a drop-in fuel and meeting recognized sustainability standards, such as EU-RED, US-RFS2 or RSB, and in particular the greenhouse gas reduction requirements contained therein' (International Air Transport Association n.d.). As we have seen, the EU has accepted the RSB as an agency in certifying fuel to EU RED standards. IATA's position is an enormously important endorsement of and commitment to sustainable fuel. But perhaps even more illustratively, IATA is a founding member of the Air Transport Action Group.

Air Transport Action Group (ATAG)

ATAG, like IATA, takes on a broader portfolio of issues than just the environment or sustainability. Further, it distinguishes itself by pursuing diverse membership: in addition to airlines and other air industry corporations, it stretches beyond individual companies of any sort and into the realm of associations and then umbrella federations of associations. However, like every group here, (and remarkably, given its very broad constituency) it is highly involved in the sustainable aviation fuel file. In fact a glance at the ATAG website (www.atag.org/) leaves one with the impression that this very comprehensive group has sustainable aviation as its principal preoccupation. For example, it funds and operates the www.enviroaero.org initiative.

If one examines the membership list, it becomes difficult to find any important player in the arena who is not somehow captured by the collective sustainable fuel aspiration. I list its members to provide a sense of their variety, and to allow an exploration of 'affiliation trees' that flow particularly from the inclusion of associations and organizations, and then federations and groups that represent large member associations in turn. A good example of this last feature is the International Coordinating Council of Aerospace Industries' Associations (ICCAIA), which comprises several separate industry associations around the world and, consequently, an extremely comprehensive representation of aerospace companies.

Funding Members:
- Aerospace manufacturers: Airbus, ATR, Boeing, Bombardier, CFM, Embraer, GE Aviation, Pratt & Whitney, Rolls Royce, Safran.
- Organizations and associations: Airports Council International (ACI), Civil Air Navigation Services Organization (CANSO), IATA
- Fuel technology manufacturers: Honeywell.

Associate Members (all associations):
- Aerospace Industries Association of America (AIA), Airlines for America (A4A), European Business Aviation Association (EBAA)

Active Members:
- Service providers: Aéroports de Montréal, Aéroport de Paris, Airservices Australia, Doha International Airport, Genève Aéroport, Manchester Airport, Romanian Air Traffic Services Administration, Sabre Airline Solutions, SITA, Società SpA Esercizi Aeroportuali
- Organizations and Associations: Aérosuisse, Arab Air Carriers Organization, Association of Asian Pacific Airlines, Association of European Airlines, European Regions Airline Association, Fédération Nationale de l'Aviation Marchande, International Air Carrier Association, International Federation of Air Line Pilots' Associations, Latin American and Caribbean Air Transport Association, Union des Aéroports Français
- Airline: FedEx
- Educational institution: Politecnico di Milano

Partnership Members:
- Organizations and Associations: International Air Rail Association, International Chamber of Commerce, International Coordinating Council of Aerospace Industries' Associations, Pacific Asia Travel Association, World Travel and Tourism Council.

I will say that airlines, related air industries, and organizations of other types all share the very real goal of figuring out a way to shift air travel to credibly certified, sustainable, alternative drop-in jet fuel. Many of the questions about how best to do that are difficult, no argument. We begin to see why when we start to examine some the on-the-ground research, commercialization, policy, and collaboration initiatives.

The Commercial Aviation Alternative Fuels Initiative (CAAFI)

CAAFI is a US government/industry alternative fuel collaboration that asserts itself a great deal. It has been a prominent part of the effort to generate interest, meet aviation technical fuel specification milestones, and build the dialogue on the commercialization of alternative sources of aviation fuel. It has been instrumental in shepherding the alternative fuel effort along. This includes building bridges between producers and potential customers. CAAFI is a major clearinghouse for corporate interests of aircraft and engine manufacturers, fuel producers, and aircraft operators together with government regulators and fuel certification bodies. It is a hub of commercial and technical interconnectedness.

As an enabler, resource, and collaborative facility, the organization largely refrains from prescription. This includes the thorny matter of sustainability standards. The case has been made that fuel production must embody the three measures of sustainability, including all of their respective sub-parts. Within CAAFI, we find a focus on environment, commercial viability, and security of supply. CAAFI's website (www.caafi.org) presents this picture pretty unambiguously, pointing out that other considerations might eventually have to be included in its comprehension of sustainability, but they are consciously being set aside in favor of those things that might more immediately preoccupy a producer (perhaps thinking most about a US producer.)

In its sustainability guidance document, CAAFI mentions the RSB more than once and provides a good summary list of other standards that might have to apply to a producer's fuel. The *CAAFI Alternative Jet Fuel Environmental Sustainability Overview* says:

> As the aviation community seeks to adapt to the changing energy landscape and facilitate the development and use of alternative jet fuels, the industry will need to ensure that *the fuels into which it invests political and economic capital* will provide the hoped-for benefits (environmental, economic, and *otherwise*). (Commercial Aviation Alternative Fuels Initiative 2013, my italics)

The 'otherwise' draws our attention inasmuch as it is unspecific about, for example, social justice elements. The guidance document goes on to talk about such things as carbon emissions, energy use, water use, land use, and biodiversity. And then later it says: 'This is not intended to imply that other environmental, social, and economic indicators are not important, and other indicators may be added to this document over time.' It is just that 'over time' is a bad time to add new, more difficult, and perhaps more expensive criteria.

CAAFI's most immediate (American) regulatory environment does not really specify much in terms of advanced comprehensive sustainability standards. This reflects the realities of the policy context for both the US organizations that CAAFI hopes to help, and the regulators under whose watch they perform, and with whom they are connected. When I spoke with Lourdes Maurice, the Executive Director for Environment and Energy at the US Federal Aviation Administration (FAA), I asked her specifically about the idea of adopting common or harmonized standards (such as the RSB's); she thought that the US position would initially be to establish national standards that could be harmonized. She expressed the concern that one jurisdiction might require a higher or lower carbon reduction than another for valid reasons, but this would create issues for producers and users.

One interviewee worried that committing to a 'highest common denominator', more universal standard might defeat efforts to accomplish important but incremental progress. In this context, it is important to remember, as was discussed earlier, that there is some difference between how we generally understand the term 'standard' and what a standards organization is able to offer. Recall that the RSB, for example is capable (at least in the CO_2 dimension) of certifying to degree as opposed to a binary determination of absolute compliance. For example, it can be said that a fuel represents a 50 percent reduction in carbon on an LCA basis, and that might be the minimum value with which the RSB wishes to be associated. But it is certainly possible for the RSB to adjust its methods and certify as to the actual carbon content of *any* fuel. Airlines have to be able to board fuel all over the world, and someone has to be able to say something definitive about how well such fuels comply with whatever standard is in place in any given jurisdiction.

The same CAAFI document mentions that in the US, fuel makers would be referring to the Renewable Fuel Standard (RFS2) but it also points out that in other jurisdictions it would be different. The EU and required compliance with its RED are mentioned, and also that fuel makers must comply with the RED through compliance with standards prescribed by organizations such as the RSB that have been approved as EU certifying bodies. It is important to note that the RSB goes way beyond the relatively simple environmental criteria that CAAFI suggests are most immediately pertinent: carbon emissions, water use, land use, and biodiversity.

CAAFI's *Overview* also says: 'it is critical to select an appropriate performance evaluation tool based on the objectives of the analysis' (Commercial Aviation Alternative Fuels Initiative 2013). That is absolutely true. But everyone in the industry should expect that all of this could go, should go, and (most practically for those who want to ensure that they reap the benefits of real and really perceived sustainable fuel use) *will* go in the direction of highest common denominator. A player in this game (such as an airline) might well wish to insist that a prospective fuel provider would be able to see its way to meeting the highest standard regardless of what it wants to demand of itself in the interim and regardless of how much or how little is demanded of it by a particular government.

CAAFI certainly seems to offer an opportunity to read between its own (and specifically US) lines: having limited the scope of its *Overview* document, it does include an excellent summary of standards frameworks at page 11 that might apply at some time or in other places.

In any event, CAAFI has been singularly effective in ensuring that the US user community is in good communication with the developers of new fuel technologies and those who might be in a position to produce, as well as regulators and organizations moving the technical suitability agenda, such as ASTM International (formerly American Society for Testing and Materials) in the US and DSTAN (the standards body associated with the United Kingdom's Ministry of Defence). CAAFI has been more effective than almost any other organization in actually helping emergence of commercially viable, sustainable jet fuel.

Commercial Voices: Producers

It is both sobering and encouraging to hear what people in the producer community have to say. We have companies that are pushing ahead and confident in their success at the same time that virtually everyone worries about the success of the larger challenge. Let us meet two of those people for now.

Jennifer Holmgren

Jennifer Holmgren has played major roles in bringing two different alternative fuel ideas to both technological maturity and commercial viability. In her position at Honeywell's UOP and now as Chief Executive Officer (CEO) of LanzaTech, she has become well versed in all facets of the quest for sustainable, alternative fuels. We have seen that she and her company have been involved in some key air industry initiatives. In conversation, and in examining her career, she seems the embodiment of the idea that a corporation must serve large purposes; that profitability is a necessary beginning to a broader mission.

When we spoke, I first asked Holmgren what she thought of the achievability of the requirements laid out to us by the United Nations' Ban Ki-moon, and the Organisation for Economic Co-operation and Development's Angel Gurria. Can we get carbon emissions to very low levels in the twenty-first century? 'No. I don't think we're on that path ... I don't think we've shown the commitment.' Holmgren is well situated, operating a business that has some bright prospects, and not having relied very much on government handouts. LanzaTech is quite a star right now with a very useful technology, partnered with big companies, and with a commercial demonstration project that has received RSB certification. Nevertheless, despite how well she and her company are doing, she similarly points out the world's need for *lots* of fuel producers using a variety of technologies, and at commercial readiness. In her mind, the stew of start-ups with their proposals and pilot projects constitutes the necessary resource, which must then provide the candidates that can be brought to commercial readiness: 'We don't know what will work.'

When I asked Holmgren about policy that would create capacities through public procurement, she replied that this would be part of the solution. She talked about the US Department of Defense procurement program as being a help, but she pointed out how complex the task of de-risking new fuel projects is. Regardless of offtake agreements, and regardless of their scale, projects still need financing, and for investors, a purchase agreement only covers one risk. There is still the risk associated with verifying the technology itself and verifying its technical scalability. In a related aspect, she described how de-risking even the financing leaves holes: government loan guarantees can lower the cost of borrowing or attracting investment, but they cannot *ensure* success in securing capital, for the same reason that offtake agreements cannot: other risks. Her view is that the full arsenal of public support policies must be applied (intelligently, selectively, and appropriately, it goes without saying) to the task of fostering the required large number of candidate enterprises required.

LanzaTech's process can use a variety of feedstocks, including wastes and residues from industry that are not 'biological'. In this sense, LanzaTech is one of those companies, about which we spoke in an earlier chapter, that do not produce 'biofuel' from biomass. A proprietary microbe is used in the process, so there is a biological element, but the feedstock—the stuff that the fuel is made from—is not necessarily biological at all. This reminds us that the words that we use in discussing or defining sustainable fuels and formulating policy are not capturing the full picture. Note again that their product has been certified by the RSB.

In the commercial demonstration project in China, the feedstock is flue gas from a steel mill. So the difference here is that (remember the importance of carbon-producing waste) the company that owns the steel mill itself, if it has a desire to mitigate emissions, can become the source of required investment capital. This reduces the need to find so much venture capital or borrowed money. There are so many good ideas that do not enjoy that particular advantage. Holmgren speaks credibly on the need for good policy assistance: that is perhaps because she has not needed so much of it, and appreciates the plight of those who do.

When our discussion migrated away from policy and more in the direction of the technologies themselves and what seems to help and what does not, there were many valuable insights. A key part of our discussion was the matter of how scalability of technology is critical to its practicality and adoption. LanzaTech's approach to seeking out opportunities to exploit its technology made it clear that the scalability matter offers two aspects: how easily the technology actually scales, but also how practical it is at scales that are not perhaps huge. Holmgren feels that it is important to be able to exploit as many niches as possible in terms of local scale of feedstock production. Operations that are feasible at smaller scales are easier to finance and generally easier and quicker to start up. This complements what we discussed, generally, about scale earlier. Systems that can operate at small scales are also perhaps more amenable to being set up in smaller markets in developing countries, where current energy demands remain relatively modest. Assuming ready supplies of feedstock (in LanzaTech's current model, flue gas) Holmgren suggests that 30 million gallons of annual production capacity might be established for perhaps $40 million instead of initial layouts on the order of hundreds of millions of dollars, as is not uncommon. At such scales, we could hope that it would also be easier to understand and control for other effects on local resources and communities.

As we have seen, there is an interesting discussion on which to engage related solely to the matters of using (a) waste for fuel production, and (b) using it as close to the point of consumption as possible. LanzaTech's current focus on steel plant waste gas fits this profile ideally: if output were to be used for aviation fuel, steel plants are unlikely to be located very far from airports.

But the technology prompts a question: Flue gas that results from the consumption of a fossil resource—whether it is the use of coal as a steelmaking chemical reagent or as fuel for energy—is still released, ultimately. The technology being discussed uses the carbon twice: once by the (steel) factory, and then again for conversion into fuel. But the fuel end product does get burned. So I was interested in whether Holmgren foresaw that as carbon-reduction targets became more ambitious, exhaust gas exploitation technology could be shifted to situations where the flue gas that was being captured and processed into fuels could be coming from renewable sources. The answer was that it could. The reason that this question is so important in the context of fuel technologies that exploit waste products is that it speaks to a possibility that we raised much earlier. If carbon-neutrality of energy is the brass ring, what we might call 'carbon-negativity' goes into the realm of dream ambitions. But timescales of several decades are big timescales, and scales of ambition should be commensurate. And (again the old saw) we should know where we might be able to go and how we might get there in order to decide at every fork in the path. Using perfectly renewable and sustainable sources of feedstock to produce energy gives us a carbon-neutral circular package. But if we are able to use the waste from such a process one more time before we release it, that offers the prospect of making a double dent in our total emissions.

It so happened that at this point in our talk (and knowing that Holmgren was not as directly involved in the discussion about aviation fuel as she had once been when working with Honeywell's UOP) it occurred to me to ask her what she thought about the long ongoing discussions at the ICAO, which, subsequent to her active participation in ICAO alternative fuel conferences and the like, had substantially devolved into a preoccupation with MBMs. She offered the insight that it would be a shame if our focus on ways of accounting for the carbon that would be produced monopolized our attention, and took away from our ability to see through to ways of eliminating the carbon so that it did not *have* to be accounted for.

Dirk Kronemeijer

There is one person interviewed here whose company acts as a link between the fuel producer and user communities. Several years ago, Dirk Kronemeijer was Vice President, Business Innovation at KLM Royal Dutch Airlines, one component operating unit of Air France–KLM. KLM (founded as Koninklijke Luchtvaart Maatschappij) has the distinction of being the oldest airline in the world operating under its original name, having been created nearly a century ago, in 1919. Capacity to innovate can be assumed to be a key asset of any company that persists for so long in an industry that has evolved and been remade by technology, scope of operation, regulatory environment, and changing markets. I spoke with Kronemeijer to find out more about how a different sort of fuel company got formed.

Several years ago, in 2006, his idea was that sustainable fuel was going to become critical to the whole industry. He investigated technologies and found out that such fuels were technically feasible, but he and others felt that not enough was being done to ensure that they would be commercially available. There was support for his idea within KLM that the airline should start to incorporate sustainable fuels into its stream. As a result, he became involved in discussions about more sustainable fuel in which major fuel suppliers were (effectively) asked the question, 'Are you going to make this stuff?' Generally, the answer was 'No'. He says that the discussions reflected the fact that fuel producers did not see any reason to cater for airlines' desires to have a different kind of fuel because they were a captive market—they had to buy fuel and would have to buy whatever the providers made. But another point that they seemed to be making to him was this: *If* hydrocarbon molecules that were sourced sustainably were to be made available, they would be more usefully (from the point of view of a fuel company's strategy) made available to transportation modes that (a) were subject to influences like mandates, and (b) *could* explore other options; if car and truck buyers and makers had the option of converting to electrical power, for example, then fuel providers might wish to offer them sustainable fuel so that they did not.

This ends up making aviation's particular pursuit of sustainable fuel a fight against both market logic *and* policy. So, a very important point is that lack of sustainable aviation fuel policy at any scale (international, national, or even regional or municipal) is denying aviation both incentive *and* capacity to change, relative to other modes, even though other modes can change *more easily*. This exacerbates what is being called the 'competition for the molecule' conundrum.

Faced with these facts and other considerations, Kronemeijer became convinced that in order for *any* airline to have access to novel and sustainable sources of jet fuel, the best way to proceed was to see that it was widely available and adopted by lots of airlines. That was the idea behind SkyNRG of which Kronemeijer became the Managing Director. The company was a joint venture of KLM, North Sea Group (later Argos Energies), and Spring Associates—(respectively) an airline, a fuel manufacturer, and a sustainability consultancy. The thrust of SkyNRG is to be a 'market maker'.

To expand the market, SkyNRG sells not only to airlines, but also to other corporations that want to reduce the impact of the commercial flight transportation that they buy. Much like buying green electricity from a distributor that sources some of its power from sustainable sources, non-airline client companies can contract with SkyNRG to pay for an amount of fuel that covers some or all of the flying that they do. This is an alternative to buying offsets for a corporation's share of the world's aviation emissions.

But it requires rigor in order to be considered credible. Kronemeijer stated that a preoccupation is to ensure that what they and their customers are doing cannot be challenged in terms of the improvements in sustainability that are being claimed. He supports RSB and regards it as the best certification standard that is widely available, but SkyNRG maintains an additional internal 'sustainability board' that applies its own rigorous standard. Credibility is deemed absolutely vital.

SkyNRG is still in early stages and needs growth. In terms of commercial design for the longer term, the company has adopted a 'bioport strategy'. This means that the focus should be on picking a location and establishing a sustainable fuel presence there. The strategy generally involves or implies:

1. involvement of the major locally based airline taking perhaps 50 percent of supply at a modest price premium (at least initially); SkyNRG takes the rest and resells it
2. some government support
3. good availability of feedstock
4. access to the airport
5. a position in the supply chain.

A strategy like this is intended to be technology and feedstock agnostic; these matters depend on the site. Each airport will target feedstocks and technologies that work most easily in that particular location.

Thus far, SkyNRG has been focused upon used cooking oil (UCO) as a feedstock. It has received criticism for this strategy. Critics charge that this is a very limited resource and cannot constitute a viable supply base for expansion. Kronemeijer's answer is that UCO must be used or wasted, and that it is an excellent place to start in getting sustainable fuel established in the market. That makes sense: if UCO is easiest, it should not be passed up. He also points out that there is an enormous amount of UCO that is not being taken up as a feedstock. If 70 percent of UCO is currently collected in the US and EU, he offers the example of China where he claims that only 1 percent is collected. The organization's clear plan is to exploit other technologies and feedstocks as they come to make sense.

But, again, he wants absolute credibility for the sustainability claims that he and his clients make, and almost regardless of the actual, technical certifiability of any feedstock/technology combination, he says that SkyNRG will be '100 percent sure' to stay away from fuel sources that are likely to be heavily challenged in the court of non-governmental organizations (NGOs) and public opinion.

This point led to an interesting discussion about the credibility of certification systems. Kronemeijer considers this absolutely key to viability of certifying bodies and those who use them. He points out, for example, that such credibility is a function of the durability of assessments made previously. In that regard, we could say that it is not enough for a certification body to be able to claim that it did its absolute best according to what it knew at the time, but rather that its claims never subsequently be proved unjustified. We could take the example of palm oil. Palm oil faces a little resistance in the market, and has been the target of criticism from environmental groups and others.[3] Even if it is possible to absolutely guarantee the sustainability of a source of palm oil for biofuel, nevertheless some commercial customers will be put off by any sustainability public relations trouble that palm oil production has experienced over the years. Without speaking about any particular feedstock, Kronemeijer is not interested in uphill battles. That is a cautionary tale for those developing and applying standards *now*.

Kronemeijer feels that we can achieve aviation's sustainable fuel ambitions. He is not intimidated by the air industries' goals of carbon-neutral growth by 2020 and a 50 percent reduction in absolute emissions by 2050. Is he aware of the kind of reduction in rate of emissions that is implied by such goals? Yes. He thinks that a 50 percent reduction in real terms will mean that fuel will have to have a carbon-content reduction on the order of 80 percent to make the 2050 target, given the anticipated growth in traffic. He calls this 'easily feasible'.

But such ambition really does seem to depend on Kronemeijer's five points, and co-dependent participation by government at every level, airlines, airport authorities and those who control airport fuel infrastructure appear essential. That makes it sound tentative. On the other hand, where such collaboration is successful in making one airport a leader, it incentivizes those elsewhere to make something similar work.

Commercial Voices: Airlines

Now, we will hear from a few people in the user community.

---

3   See, for example: http://www.saynotopalmoil.com/Whats_the_issue.php; http://www.orangutans.com.au/Orangutans-Survival-Information/What-is-palm-oil-and-why-the-controversy.aspx; http://www.nature.com/news/palm-oil-boom-raises-conservation-concerns-1.10936; http://www.nytimes.com/2007/01/31/business/worldbusiness/31biofuel.html

Virgin Atlantic

It is interesting to examine the path taken by organizations that have started earliest, accomplished a lot and, presumably, learned the most. Virgin Atlantic Airways Ltd (VAA) is one such. To find out a little about its path toward sustainable fuel, I interviewed Emma Harvey, Head of Sustainability, and Jonathan Pardoe, Deputy Group Treasurer and Head of Fuel. They pointed out that in 2008, when Virgin Atlantic undertook to demonstrate the use of biofuel in the flight of a commercial aircraft, the initiative was deemed by many to be daunting. That, in itself, is important. It is essential to remember where the airline and fuel industries were just a few years ago and how far they have come in that very short time; as mentioned earlier in the book, several subsequent flight programs have demonstrated flight with alternative fuel.

One airline acting alone—even one that burns approximately 13 million barrels of jet fuel per year—is limited in its leverage. So one of the early actions that VAA took was to be part of the founding of SAFUG. We have seen a little bit about SAFUG and this connection was an important part of what I wanted to discuss with them. Pardoe was involved at that incipient stage. Harvey pointed out that those from VAA participating at the time (Pardoe and others) felt that a key reason for such involvement would be to take steps that offered some prospect of helping the air and fuel industries pursue relevant and common sustainability principles. This, of course, has been an important part of what we have noted here earlier, and it is significant to see an industry stance in favor of such high, common standards. She said that they considered it absolutely critical to the industries' purpose and credibility and the worth of the entire effort. Harvey said that probably SAFUG, Virgin and everyone else involved is sensitive to the fact that 'it's really important to stay away from the whole "unintended consequences" effect [such as resulted from] earlier generation biofuels,' pointing out the difficulty of recovering from commitment to fuel production pathways that, in the end, must be abandoned because they simply did not accomplish enough of the right things. She said that this further underlines the need for 'the most robust sustainability standards'. But, one has to ask, what would that mean unless the principles and standards have been generally agreed? As discussed earlier, many airlines support the RSB as the standard that best addresses the need for robustness, and Harvey points out that VAA is one.

The SAFUG website (www.safug.org) states that it was formed 'with support and advice from the world's leading environmental organizations such as the Natural Resources Defense Council and the Roundtable on Sustainable Biomaterials (RSB).' We can thank founders and early adopters (like VAA) for proceeding in this way because we need standards to be both demanding and uniform.

The SAFUG initiative is largely working—for itself and for the organizations, such as VAA, that undertook to create it. Airlines do not actively recruit other airlines to join but, as pointed out earlier, the airline membership represents about

32 percent of commercial flight and Harvey says that applications to join come in steadily. Pardoe claims that airlines are very interested and curious, and that the issues of sustainability, SAFUG, and bodies like the RSB and other sustainability certification standards routinely come up at industry meetings, such as the IATA fuel forums. They both pointed out that SAFUG governance rules request that new airlines joining the group be willing to participate actively and help with the work.

I wanted to know how VAA's efforts were delivering in terms of actually getting fuel in tanks. Harvey described how, by 2010, VAA was looking for ways to make its commitment to sustainable fuel more 'relevant, concrete, and commercial'. In deciding to pursue a regular, steady source of sustainable fuel for some of its fuel needs, the company faced a slightly intimidating challenge: Lots of new, early stage companies offered a variety of feedstocks and, in VAA's eyes, largely untested technologies. They generally asked for significant investments to help them work through the development stages to commercial facility, and expected to sell jet fuel at way above the petroleum-source kerosene price, which is not economically viable for an airline. VAA faced the challenge of sorting out which of the array might meet their sustainability requirements *and* make it through to commercial viability.

Parenthetically (and something thing that we hear specifically from providers like John Plaza, mentioned below) it tells us something about the risks that we run when policy does not offer the smaller players, with perhaps less mature technology and business development, a way to showcase their idea in a manner that gives assurance to potential customers (and of course investors). It presents us with the question: Are we losing good people, ideas, and enterprises? Furthermore, have these small organizations that are spinning their commercial wheels already been using government support in the form of research and development assistance such as grants and loans, and are we losing also the benefit of such public investment?

In any event, Harvey reports that Virgin Atlantic found what it wanted and has partnered with LanzaTech as a potential fuel supplier. One specific thing that made LanzaTech immediately appealing was its focus on waste stream exploitation for its feedstock. VAA had consulted with a number of environmental NGOs about the potential for low-carbon fuels and these groups were generally supportive of exploiting waste streams as the best way forward. Also, as we know, LanzaTech was happy to work on RSB certification for its demonstration project in China (since achieved) and this met a key VAA criterion, since it asks any of its potential suppliers to be RSB certified.

We have noted that LanzaTech's fuel from waste gas streams is not a 'traditional' biomaterial and Harvey pointed out to me that her impression was that RSB was very enthusiastic, flexible, and adaptable in advancing the certification efforts while applying its standard rigorously. It stressed sustainability rather than technology. This confirms what the RSB's Barbara Bramble told us earlier.

Another thing that made LanzaTech interesting according to Harvey was the very fact that it had a project in the works in China: 'Low-carbon development in a developing country.' Good point.

Virgin Atlantic seems well on its way to incorporating an incrementally increasing amount of fuel into its system. But I think that we can see that it takes effort. When I asked about the kinds of policies that might help airlines and fuel companies to achieve more, both Harvey and Pardoe admitted that it was a complex question. They reflected upon all of the policy work that we have discussed previously as being done at national, regional (such as in the EU) and international levels. But Pardoe did have a couple of comments: It is important to exploit the necessary fungibility of sustainable and more conventionally sourced fuel. The best strategies will be based on getting sustainable fuel blended with other sources as far upstream as feasible, and just treating it as a 'normal' part of the fuel supply rather than as a separated, differentiated commodity. If, as would happen ideally, the fuel is produced relatively close to an airport, it can be fed into the pipelines that serve the facility and consumed as part of the regular feed. Who made it, and who bought it (and consequently gets the carbon-reduction credit) is a matter of certificates. Who burned it generally matters not in the least. The VAA-LanzaTech deal in China would emphasize that, perhaps, if VAA were not operating there and likely, itself, to take the actual fuel in Shanghai. The point is that as the number and operating locations of potential customer airlines grows and the number of airports where more sustainable fuel might be offered also increases, it becomes absolutely necessary to just feed the sustainable fuel into the pipe at whatever point is logistically most convenient.

It is interesting that, on the policy front, one of the main points that Pardoe had, apart from supporting the kind of continued financial help that governments *have* provided to fuel developers and providers, is one thing that they should *not* do: impose quotas and mandates. That is not a new idea here either. He pointed out that it is just too constraining and bureaucratic for an industry that is developing in a dynamic way, and for airlines that want to buy from the array of their (otherwise) best of possible options. This seems quite valid and of course confirms what has already been said in earlier chapters about such measures. We can surmise that quotas and mandates are perhaps putting or keeping other options out of business. Extending the point—as Harvey and Pardoe did—we can say that there should be a level playing field in competition for the available fuel molecules. Policies that set mandates or quotas *even in other sectors* can distort fuel markets in a way that makes completely fair access almost impossible. And as other commenters have noted, such policies can contribute to driving the pursuit of sustainability in the wrong direction and effectively remove some options from the market.

GOL

In the minds of some, the low-cost carrier (LCC) is the archetype of 'real' airline commerce: attempting to build business by keeping fares low and striving for profitability by controlling costs. Depending on conditions of the market in which an airline operates, this can be a commercially challenging pursuit. Margins are slim, efficiency must be high, and luxuries are not tolerated. So, to the extent that the effects of air carbon emissions are an externality unrecognized by markets, there might be an expectation that LCCs would be less interested in addressing them than most, regarding such acknowledgement as the kind of balance sheet indulgence in which only spendthrift full-service airlines would engage. Well, it is true that LCCs would not wish to spend any more for their fuel than they have to. On the other hand, some seem willing to invest considerable time and treasure in figuring out ways of securing sustainable supplies of fuel (albeit preferably at prevailing jet fuel market rates).

I spoke with Pedro Scorza, who serves as Director of Technical Operations at GOL Linhas Aéreas Inteligentes S.A. Much of our discussion in this book has focused on the international arena and most of GOL's operation is domestic, but what Scorza has to contribute is relevant. Operating mostly in Brazil, GOL is the largest LCC in South America. Evidently it considers that its clientele values sustainability; GOL is certainly working on securing sustainable fuel for its operations. What actions has GOL taken? One was to participate in SAFUG. Scorza points out that participation in such an organization is extremely helpful from the point of view of keeping abreast of developments in other countries and at other airlines.

GOL will have some things to show its SAFUG partners: it is taking action in an interesting way. The airline has been a prime mover in an initiative called the Brazilian Biojetfuel Platform. This is another example of an airline acting on the assumption that the best way to secure a supply of sustainable fuel is to help foster a domestic industry capable of supplying it in large quantities. This pragmatic approach reflects corporate policy that emphasizes the necessity for initiatives to bear the burden of their own specific costs. Using sustainable fuel is assumed to provide benefits, but they do not fall under the aegis of departments dealing *specifically* with sustainability or corporate social responsibility (CSR); rather the case must be made and project participation managed by regular operational departments.

Are there public relations benefits? Of course. But people not so engaged might not appreciate the challenges of importing values and a sense of the larger social consciousness about things that seem intangible, like the need to shift toward sustainability. So it is important to give the matter public profile in order that the broadest part of the public is aware of the sustainability benefit. One of the most immediate payoffs of demonstration flights and projects is to raise not only public awareness but also *internal* awareness about the issues. The Brazilian Biojetfuel Platform originally grew out of a will to showcase to the larger society the benefits of moving to sustainable jet fuel, by conducting biofuel flights during the period of Brazil's hosting the World Cup. But Scorza says that one result has been to cultivate a deeper internal corporate commitment.

The Brazilian Biojetfuel Platform has advanced the agenda. That is primarily because it has provided a widening opportunity for participants at any point in the value chain. Scorza points out that, in general, roughly 85 percent of biofuel cost is in biomass and logistics, while only 15 percent goes to processing and production.

He stresses that the development of alternative, sustainable fuel supplies should be most sensitive to location of the most affordable sources of material and the location of the largest fueling points that can exploit the technologies available. This implies varying sustainable feedstock, technology, and manufacturing, according to volume and location of need. This again underlines the need for alternative fuel efforts to be open to participation of the broadest variety of partners and agnostic as to biomass and technology.

Can policy help? The project collaborators are keen for the country to realize that all of this constitutes new national economic capacity. Dollar-denominated costs in the Brazilian economy (jet fuel most relevantly for us) are on the rise as this is being written. Media reports are highlighting the significance to the air industry with GOL's fuel costs reported at approximately 41 percent of total cost, and the highest in the Americas for an airline having a fleet of 10 aircraft or more (Rabello and Gamarski 2013). With that in mind, Scorza advises that a case is being made to state governments that the state value added tax should be reduced to support biofuel blend introduction, and that federal taxes should be eliminated from the full integrated biojetfuel value chain, with the goal of growing domestic supply.

What of sustainability certification? If the nominal aim is to shift to more sustainable flight energy, the importance of security of supply and stability of price are still huge. Yet the sustainability of alternative fuels must be certified in order for that distinction to have its greatest value. Scorza points out that the RSB standard has been adopted by the Brazilian Biojetfuel Platform, since it has the two key elements: it has established a very credible starting place, and it enjoys broader support from industry, government, and NGOs than any other.

We immediately wonder how easy it is to assemble such partnerships, which involve so many stakeholders, because the ability to reproduce such initiatives elsewhere is important to the further development of global sustainable jet fuel production. I also spoke with Mike Lu, who is CEO of Curcas Diesel Brasil, President of the Brazilian Jatropha Producers Association (ABPPM), and integrator of the Brazilian Biojetfuel Platform. Lu confirmed that these are complex collaborations that require shepherding, coordination and a great deal of willingness of all stakeholders to work together to implement a highly integrated project. But such an initiative can clearly serve the interests of all participants, making them willing to cooperate effectively. And so he feels that such arrangements can definitely serve as a model and useful first step in building an industry.

Policy Snarls

If progress and action is the goal, what do we think about our earlier policy discussion in light of the comments here? There is certainly some pessimism and frustration.

To see how policies can become tangled and ineffective (as has already been said of the RFS2 in the US) I invite you to read through 'Obama Messes with the RFS', an article in *Biofuels Digest*, a popular biofuels industry online publication (Lane 2013). This piece was written by Jim Lane at the beginning of the comment period after the US Environmental Protection Agency proposed changes to biofuel mandates. Some may find it opaque. Point. This commentary offers a flavorful nugget of insight on how the desired effects of production mandates and quotas—hard for the average voter or energy sector worker to understand in the first place—can be lost through the relentless to and fro of national and global events. The result may be that they come to seem counterproductive to many of those who were the original policy's intended beneficiaries. The US is the largest Annex 1 economy and is critical to, and illustrative of, *everything* that is happening and has been done on almost every front. Remember though that the most interesting alternative fuel initiatives can crop up anywhere. So the whole discussion, including policy, should be of interest to people anywhere as well.

John Plaza

We are going to meet one more person. I leave the last contributor word in this chapter to an individual who understands all of these matters better than most: John Plaza. Plaza is President, CEO, and founder of Imperium Renewables Inc. (IRI). He is another former airline pilot. But he decided to *do* something about sustainable fuel rather than merely write about it. He took his own money, borrowed more, and started a company. IRI is now a major supplier of biodiesel—*the* major supplier in the US Northwest. John has climbed down into the policy fray in the US on more than one occasion, and has very strong views on what could work, what is not working, and what should be avoided. He has the unique advantage of being able to think about US policy in a global context; it was the international flight energy conundrum that initially intrigued him as he flew B-747 freighters across the North Pacific. He cultivates a strong interest in all aspects of sustainable fuel of all types. Though IRI is currently focused on biodiesel, there is a stated corporate interest in examining other sustainable fuel challenges broadly and deeply. The aviation flight energy question is right at the front of his thinking when he imagines other 'endeavours'.

When I spoke with Plaza, a point that came across very strongly was his view that the easiest way to change what society does is to make necessary change the least difficult thing to do. That may sound like a tautology but, on reflection, it is not. And the central point about Plaza's criticism of US policy, and the fuel energy policy errors that he sees elsewhere, is that policy tends to make things relatively easy for established interests but not so easy for those who can effect real change. For example, petroleum producers were certainly critical to meeting energy demand in the past, and so support and subsidy policies that favored their development are entrenched and politically difficult to remove, even though one national goal might be to shift toward sustainable options. He points out that biofuels, for example, are in a 'tiny, nascent stage' where they absolutely need policy support in the form of subsidy to become commercially viable, but that the argument on the other side is that they only make sense if they are commercially viable. The question becomes: If we refuse to withdraw subsidy for the 'traditional' fuel providers while we also refuse to help alternative sources, how is all of the resulting difficulty making it easier for new fuel producers and potential buyers to change?

As we have noted elsewhere, where policy does recognize need for change, it often supports organizations that already enjoy a relatively prominent place in the 'alternatives' portfolio. Market entry is most difficult for the smaller actors who have the most novel technologies. I hypothesized in Chapter 13 about public procurement being used as an effective policy tool. Plaza has long been a proponent of this general idea. But I think that one should also remark that where it *has* been used, such as in US military procurement programs, the technology and volume provisions tend to rule out the participation of small producers. This is exactly the opposite of what policy should do, which is to give a chance to the potentially best, regardless of current financial strength.

Remember that we have said that change must not be just a long-term goal, it is an intrinsic part of the goal that it happen quickly. Plaza offered an insightful comment: 'Policy does not go to the core of where the renewable fuel industry is.' I like the perspective that such an observation offers: the policy-maker *must* examine the undertaking from the point of view of the entities that want to make and use the best fuel. The *realpolitik* of dealing with the established interests comes after. Doing it the other way around, if we acknowledge that it is politically pragmatic, *does not* engender rapid change. In Plaza's view, the choices are becoming stark: help the sustainable fuels industry, or tax the production or consumption of fossil fuel. He reinforces the point made many times here that aviation offers the easiest and most efficient policy entry to the liquid fuels problem. Even if policy leaves sustainable fuel without a downward final price adjustment, sustainable blend components can be introduced with a very modest effect on price at the pump. In general, Plaza hopes for what might be called a 'small is good' policy perspective. Policy must ensure large numbers of potential producers (and consequently, many small ones) in order to permit sufficient capacity for innovation.

He made one other point that, while I think that it should probably be obvious, is not well understood or internalized by the political establishment, media, advocacy groups (for *or* against) and those who toil in the policy mills: Public attitudes are crucial and should be informed, unambiguously, by the clearest and most honest representation of the relevant facts and our situation. We can see in some media reporting that advocacy against alternatives can seriously hamper the development of clear public understanding. Shortly after an alternative fuel demonstration flight took place, an article in *The Australian* reported that environmental groups were calling the biofuel flight a 'stunt' and an attempt to divert attention from an irresponsible attitude to climate change, and also disputing the airline's choice of biofuel (*The Australian* 2008). What is important is what is not said: airline, fuel, and other industries were already interested in proper sustainability criteria. So the article's (un-emphasized) points about possible better feedstocks were not elaborated with information concerning sustainability certification efforts already under way at that time, even though such efforts involved participation by some environmental advocacy organizations. Policy formulation has to see through public relations fog. To that end, public communications about what makes a human activity sustainable and how that is measured is an essential precondition to policy that supports it. The two complementary things that it is essential to ensure in this regard are: (1) no false claim to sustainability should be possible, and (2) every legitimate claim to sustainability should comprehensible and credible.

So Where Does That Leave Us?

Thoughts presented here, from some of our society's most insightful observers on matters of sustainable, portable liquid fuels will probably leave some people both encouraged and perplexed. That is certainly the way I feel. With so much that offers itself in terms of new technological tools, new ways of understanding the challenge, and many enthusiastic individuals in both the airline and fuel communities, why is it that we cannot paint the picture? The complexity seems to lie in accessing the potential solutions contextually and appropriately, allowing us to figure out what to do first, where, next, after and so on. In many respects, it is not the problem that is too complicated, it is the solution.

In the next, and last, chapter we will sum up the issues that we have discussed and see if there are some things that we can do in order to foster a 'culture of solution'. Thinking more specifically about the aviation industry, what steps do airlines and potential fuel producers have to take in order to create a practical way of allowing their own particular suite of solutions to gain traction and advance. In the face of seeming paralysis in the talks that take place as between governments, how can aviation give itself an effective voice?

Chapter 15

Conclusion: There is No Such Thing as 'Business As Usual'

Time is slipping away, and yet patience is required. We must certainly accept that the world faces a decades-long commitment to fossil fuels; nothing, in any foreseeable course of events, can suddenly make that different, and that is particularly true for commercial aviation. But we do have to start and rapidly accelerate change. Where is the effective campaign to head off disaster in the longer term through change?

Are We Stuck?

Among all of the things that one can see when in Los Angeles, there is one that attracts relatively little attention. It is a startling display of the way events can unfold, and an absorbing example of a peculiar natural phenomenon. Situated right in the middle of the city, on Wilshire Boulevard, the La Brea Tar Pits are not, in fact, 'tar pits' at all. They are the accumulating residue of 'oil seeps', places where a thick, heavy, sticky form of crude oil oozes to the surface of the Earth of its own accord. The oil fractions that evaporate more easily dissipate, leaving a tarry mess. For a very long time, people have been discovering with fascination the creatures that met their end in these shallow deposits of petroleum. The preserved carcasses of enormous animals—some of the species have been extinct for centuries—have been found and studied. Contrary to what many people assume (and I assumed) these animals did not die by falling into and being immersed in the oil. They simply got stuck. The pools themselves do not look threatening, and, anyway, they can become covered by a layer of loose, wind-borne soil and dust, and not be apparent at all. For the most part, they do not start off being very deep either. But what has happened, over and over, is that the animals just unwittingly walked into a shallow puddle of the stuff and got stuck. Many became aware of their predicament while standing, in perfect health. Their life was not immediately nor directly threatened by the oil itself but rather by the consequences of being trapped in it. Unable to move, they thirsted and starved and tired. They eventually fell and lay down and died in that spot—much like flies on flypaper. All of this I learned when I visited the site with friends and we listened to the expert volunteer guides at the adjoining Page Museum.

What do we know about our own predicament? In some ways, oil has become a similar trap for us. We were not aware of any threat when we took those first steps.

And it does not threaten to kill us now. In fact its use, along with the use of other fossil fuels, has provided concentrated-energy wealth that has brought advantage, progress, and luxury for those who have exploited it. But the cheap advantage that fossil fuel confers preoccupies and transfixes us. And our situation now makes us and our descendants all victims of the effects of becoming mired in the carbon economy. The air upon which we depend for our very breath has become altered by the combustion of fossil fuels: it traps too much heat. We absolutely need that trapped heat in order to keep from freezing, but now there is too much of it. The ironies here are endless.

But even though time *is* running out, and even though we know it, we feel that our capacities to meet and deal with this new challenge are stretched. I cannot help but remember the first *Toy Story* movie (1995), and Sherriff Woody's response to Buzz Lightyear when, in a bad situation, Buzz advises that it is no time to panic. Woody objects: 'This is a perfect time to panic!' Should we panic? No. But we ought to act very quickly in an intelligent way. Are we being quick and smart? The world community of nations hopes to come to agreement on emissions limits in 2015. That is, by any objective standard, very slow (assuming that such agreement were in fact to be reached).

Achieving global agreement on emissions limits does not, in any event, constitute agreement on how to meet them. Of particular interest to us is that biofuels and other alternative fuels technologies may be asked to play a major role in providing necessary supplies of portable liquid energy. However, far from having some agreement on how to go about putting such fuels in place, there is not even international agreement on sustainability criteria for their production. Just that one fact constitutes an immense challenge about which the world has been able to decide almost nothing (Bastos Lima and Gupta 2013).

I have suggested in these pages that perhaps commercial aviation could serve as an example and sparkplug. But considering efforts at the international level, deliberations at ICAO will not produce even a *plan* to reduce international aviation emissions before *2016*. And that plan will mostly be about accounting for and securing credits for carbon emissions. There is very little scope at the international level for actions or policies that help with the actual development and deployment of sustainable alternative fuels in a targeted and specific manner. And even agreement on restricting emissions and accounting for them in a reliable and disciplined way does not really help with the underlying need to make those sources of emissions go away over the next few decades.

From Whence Action?

If the air industry (or any industry) perceives a need to adopt and demonstrate sustainability as a key criterion of its business practice, it is simply not useful to allow a vacuum in governance to dictate action and outcome—or lack thereof. Challenge, in isolation, is not necessarily and solely the province of supposedly responsible authority; it belongs to those who recognize it. The really positive aspect of what we have been examining in this book is that the industry itself wants to get on with the job of achieving sustainability.

Urgency is clearly not effecting sufficient action. Why? The air industries know very well how clear the climate change issue is becoming in the minds of those people who use their services. Academics and commentators may argue about how much coverage the matter is currently receiving (Hope 2014), but this misses the larger point. Regardless of the specific degree to which media are reporting on global warming, climate change, or phenomena related to them, the thousands of news stories over the years and decades have had their effect: People do know and understand that there is a problem. It is hard to imagine that very many who currently use air transport are ignorant of the issue or the profile that the story has gained. They ask themselves what these companies are doing. All of this has built some anxiety in corner offices, so airlines and aircraft manufacturers, and a few fuel technology companies, do want to act. In fact they *are* acting—some of them, to some degree. The problem is that though we are getting there, it is not happening nearly fast enough.

It is probably time to make some assumptions: We should accept that humanity has already committed the world to major climatic effects. And many people probably think that we have allowed the fate of the future world, with its freight of humanity and all of the life forms that interdepend, to slide over a precipice. Many of those people fly, or would like to fly. And they may wonder: Will the air industry extend one slippery hand to reach out, grab, hold, and hope that the world can find a new footing on that hostile cliff face? Or is it rather an industry that embodies a growing sense of failure and foregone conclusion? Is aviation answering the bell, or throwing in the towel? Many voices in the industry, including those organizations that speak for it, have made it clear that the desire is to get busy with solutions. There is no purpose served in contemplation of failure. Failure—famously—is not an option. The job of airlines, aircraft manufacturers, and novel fuel-makers has become clearer: If action is not enough in the face of policy that makes action financially futile, then an assault must be mounted on the bits of public and political ignorance and myopia that coddle the proponents of failure—those individuals and organizations that nurture an unhealthy stasis.

This book was intended to show aviation's place in the hazard that humanity creates and aviation's part in extending the hand. Success will lie in determination and action. The world's growing heat burden, ocean acidification and other effects will start to erode our capacity to react in a socially coherent and economically useful way. The same people who hear about global warming also hear, from those who want to hold back, that proceeding at a quick pace to bring rates of emission down is not economically realistic. But they probably also recognize with increasing clarity that it is economically unrealistic to go so slowly. Now is the time to engage. Not only must industry engage on the task of sustainability itself, but it must also engage the public and our national and local political communities. This should be a story. Not a story about failure and catastrophe, but about how, and how much, we can all win.

But the task will be great. The first things that must be done are to both face facts and then frame those same facts in a way that does not defeat our will. If we must act now, how much do we have to do? 'As much as possible' has no real resonance; it sounds 'unscientific'. We discussed earlier the '50-50-50' challenge advanced by Ban Ki-Moon. It is an unfortunate commonplace idea in some quarters that the United Nations is sometimes less than perfectly realistic and effective in its approach to issues. So perhaps some will look askance at Ban's articulation of the predicament. They will say that it unnecessarily dramatizes the plight and overstates what needs to be done—but the opposite is true. In October of 2013, Angel Gurria, Secretary General of the Organisation for Economic Co-operation and Development (OECD) made a speech (Gurria 2013) in which he references an OECD paper, *Climate and Carbon: Aligning Prices and Policies*, that urges governments to take a very hard look at their policies concerning fossil fuels, and their ability to meet climate targets (Organisation for Economic Co-operation and Development 2013). The overall message in Gurria's speech goes *beyond* Ban's plea. He states flatly that the world must not only reduce emissions dramatically over the next few decades, but that this reduction *must* progress to effectively zero global emissions. It should be clear that arresting global warming and climate change is no longer an abstracted 'environmentalist' ideal. It is the way that human society and its economy will survive and advance.

The International Energy Agency (IEA) makes available a data summary of its global analysis of energy issues. Forgetting a business-as-usual case for a moment, the 2013 IEA summary describes the following: in a scenario where all policies under *current consideration* are enacted, total, global primary energy supply (TPES) predictions from 2020 out to 2035 would consistently exceed the amounts required to keep global warming below the 'acceptable' upper limit of approximately 2° Celsius (called the '450 scenario') (International Energy Agency 2013). This would imply that in order to achieve a reasonable chance of keeping human society and its economy ticking along, we have to come up with *many more* and *better* policy ideas than we now even contemplate, and then carry through on every single one of them. These are not Cassandra analyses. They come from extremely credible organizations that are focused in a hard-nosed way on economic realities. And some might say that such commentary *still* understates the problem. I cannot imagine very many reasonable critics saying that they *overstate* it.

All of that certainly makes our situation seem intractable—intimidating. How do we frame such things in a way that speaks to our desire to win, as opposed to an oppressive foreboding about seemingly inevitable defeat? What is the great economic response in the face of such material challenge? That we should give up? Go slowly? That is not acceptable. We simply cannot be so stunned by the size of the test that we abandon the task. If industry and government policy commit to action, we will learn as we proceed and the job will get easier as we go along and as individuals and companies, and whole societies, start to profit from our advance.

One obstacle will be to ensure that the benefit from our advance into such things as new sources of energy will benefit everyone. Fairness is an essential element of any possible success. Think about our goal of more or less zero emissions. Obviously, zero means zero (eventually) for everyone. Fairness and equity do not mean unconditional license for those who struggle. No sector, however important, and no country, no matter how disadvantaged, gets a pass in the long run; barring some new, miraculous way of drawing carbon *out of* the air, there is ultimately no way of offsetting failure. This is precisely why we said earlier that the provisions of Common but Differentiated Responsibilities (CBDR) and the like must sunset. Interim obligations can be different, but plotting out the time and carbon-reduction levels of any plan (not just commercial aviation's) that shows goals like carbon-neutral growth in the near to mid-term (say 2020), and large reductions in a few decades (like a 50 percent reduction by 2050), implies that our capacity to reduce emissions must be well established in the near term, and that we see a truly spectacular downward trend by mid-century. We have to recognize how this underlines the urgency for developed states and emissions-intensive activities of relatively wealthy people to take the technology challenges very seriously—*and act to address them now*. Because the technologies must come from somewhere. If they do not arise spontaneously in less wealthy nations (generally unlikely) they must come from the current high-emitter nations with the economic capacity to work effectively on them—developed nations, rapidly developing nations, and the sectors (like aviation) that exploit energy wealth most extravagantly.

For developing nations, the curve will not be (again, notionally) carbon neutral by 2020 with a 50 percent reduction by 2050. Their curve will be large increase for the next few decades, followed by what should be an especially dramatic plunge toward zero at some point early in the second half of the century. These countries will not be able to do that unless the technology is in place, and it will not be in place unless relatively wealthier countries and economic sectors put it there. So, the reality is that aviation is a part of that world that must aggressively attack emissions where it realistically can, and then move the solutions throughout the flying world—and do so very quickly.

But we have to get analysis right and it has to take the longest view. Of consequence for the air industries, the single most effective policy, from Gurria's point of view is to establish 'a big, fat price on carbon' (ITN 2013). His argument, supported by Lee, Lim, and Owen (2013a) and others, is that pricing carbon is the most *immediately* useful device for emissions reduction. That is true, but not subtle. As we discussed, uptake of technologies is critical. Associating a cost to the emission of carbon (through cap and trade or offset purchase) will drive carbon use down as rapidly as we want, if emissions-reduction tools are in place. But a simplistic view that a price can be assigned to carbon emission, and then gradually migrated into the developing world in the latter part of the twenty-first century, forgets that if carbon-reduction technologies are not available in those countries, they will basically need to shut down their economies. That will not happen without a destructive fight. And it should not happen at all. Furthermore, it *need* not happen. The goal of policy can and must be to make the necessary tools available to everyone. Simply pricing carbon and blindly prosecuting that as the matter of the highest priority cannot accomplish what we need. Considered in isolation, the policy tools that encourage development and uptake of new energy technology may not immediately produce the same level of carbon reduction as just charging for its emission—the OECD report makes that point—but development and uptake *are* the essential first (unfunded for many countries) steps that allow carbon pricing to be effective.

This book has spoken, in many places, in terms such as 'aviation should do this and such', but airlines and manufacturers of airplanes and engines do not make fuel. How can the industry develop a role for itself in the conversion? It is not in the interest of aviation petroleum fuel providers to remove carbon from the fuel. In fact, carbon is the thing that they *have* to sell. Not only that, but they have many more customers than just airlines to sell it to, as well. And they have enormous amounts of capital tied up in sourcing larger stores of carbon in new and more expensive ways. As the OECD's Secretary General Gurria noted in his October 2013 speech, governments also have a strong stake in the carbon market; he referred to this as 'carbon entanglement' (Gurria 2013). That is one consideration. Also, aviation does not have equipment suppliers that can take carbon *out* of their exhaust streams. Aircraft and engine manufacturers can only help to reduce the amount. And, again, an airline has absolutely no internal capacity for developing a new fuel. It can only offer to buy better fuel. Given relatively small margins, it can only afford to make that offer if the fuel can be provided at pretty close to open market rates for energy, including conventional fossil fuel sources.

Some relatively wealthy airlines in some relatively wealthy or rapidly developing countries can take some steps on their own to stimulate alternative sources of fuel for themselves. But it is beyond the wherewithal of even the wealthiest to undertake the project of ensuring that similar capacity is created for other airlines around the world, rich and poor. So it is important that policy-makers, and the industry entire, recognize that incentivizing new, more sustainable fuel (and other) technologies is an essential piece of the puzzle. If incentives based upon funding research, funding pilot projects, funding development, and funding deployment do not work (to a satisfactory degree and quickly enough) something else *must* be *made* to work. Actually, everything depends upon establishing that capacity. Nothing really works without it.

But air travel, with a world-scale task, has only national, regional, and local governments to whom it can appeal for help. If the only actors of consequence are found at national and lesser scale, how does such an evolution proceed? For example, how does 'international aviation' lobby the governments of developed (and rapidly developing) countries to do the things that are useful? How does an individual airline do it? There was a time, before the Chicago Convention, when airlines undertook everything themselves, as regards interacting with governments. If they needed something (a right or a facility for example) they dealt directly with whatever government might be involved, whether their own or another. Those days are gone. Governments wanted them gone. But now, it seems, we need airlines and fuel companies, in partnerships or independently, together or individually, to be able to interact not only with the governments of the countries where they are based, but also with those of countries with whom they might cooperate in establishing the commercial viability of some amount of fuel production capacity.

The job seems to require, simultaneously, each airline and all airlines acting together to address multiple governments, at multiple levels, and make a case for changing a lot of policies that currently foster great failure for the many but great advantage for some influential people and corporations. States that can do something to effect change must be convinced of the need to act. If United Nations Framework Convention on Climate Change (UNFCCC) principles are valid and still in place—and they are—the only recourse of Annex 1 countries and rapidly developing countries is to ensure that sustainable alternative fuel capacity is developed at home and then also shifted to those locales where their airlines and other airlines need it. None of the wealthy countries balked at taking the original fossil fuel wealth advantage.

The key is this: It is true that when we frame the challenge as I just have, it may seem impossible. But we have to remember the characteristics of policy that we would like adopted. Immediately, it then becomes apparent that we only need *some* national governments on side with *some* more local governments, involving *some* airlines and their fuel partners. Progress then incentivizes progress elsewhere and provides the technological capacities that can migrate into areas where there is no independent ability to put them in place.

We are talking about flying, in this book. If the air travel industry collectively is able to start an evolution toward sustainability, *that* is how and why it will happen for others. Because the same things apply to any energy-intensive or otherwise carbon-emissions-intensive activity: surface transportation, manufacturing, power generation, making cement, clearing forests, industrial agriculture, or any other such pursuits. We have three choices as regards the services that we enjoy as a result of these activities—and three choices only:

1. Make the service sustainable (including getting it to zero emissions)
2. Abandon the service
3. Change the world in enormous ways that are not comprehensible, predictable, or quantifiable.

The economic usefulness of aviation and the consequent economic usefulness of sustainable, alternative sources of fuel must translate into policy approaches that help those who are in a position to benefit commercially by *being* useful, and not those who contribute to the problem. The job of those who have the most direct interests in air travel—and that would include users of the service as well as its producers—is to point out to governments the value of the service, the value of providing it sustainably, and (very importantly) the value of what has already been spent around the world in research and development of more sustainable ways of sourcing fuel. This last point is critical: What was the point of investing billions in new fuel technologies if that effort is not followed with getting those technologies into every possible market? It won't go for free; someone will make money and pay tax.

The insight that is essential is that it is not necessary—or even possible—to get every important decision-maker to accept and support the logic of sustainable energy (or sustainable anything); it is only essential to get *enough*. If it is true that the air industries want to do this, they must get started on the task and find every single sympathetic ear in every country at every level.

It will be a frustrating campaign on some fronts but a happy bit of inevitability is that it will be very rewarding on others. Lots of people—including those in some governments—are aware of the hard economic (and newly discovered) reality: energy that creates more cost in its production than benefit in its consumption has no value. In the face of known greater harm, incredibly, some governments or corporations still argue about the local economic benefit of, say, a coal mining operation, and that any changes should proceed slowly and cautiously. That is really just arguing that concentrating benefit immediately and locally is acceptable if the cost is suffered more globally. This is a kind of collective sociopathy. Not all governments are that myopic.

Failing

Air transport consultant and commentator Chris Lyle pointed out to me that he thought there may have been a diminution in the amount of optimism that had, until recently, been tied to the effort to develop sustainable, alternative sources of jet fuel. This may be true. And remarkably so, since the only thing that has actually changed over the last few years is our collective estimate of how prospects for new technologies are improving. The technical challenges are largely solved. That is something that is routinely forgotten. We do actually have ways of making better fuel and producers who are capable of doing it. We are being defeated solely by our political challenges. In fact, even our successes are conspiring to frustrate us: Now that forward-thinking airlines are starting to talk about the real feasibility of sustainable fuel and even contracting to buy some, many think that the problem needs no more attention. But it is not only the fact of inevitable change that is important, *speed is critical*. We need intelligent, rational policy action to allow change to proceed at the necessary fast pace. For every single problem that we have in the world, the rate at which we expand our powers of solving it has to exceed the rate at which it is growing, or it will defeat us. But, again, there are lots of people who understand that, even if it seems desperately hard.

Of course it helps to be able to say encouraging things to move the case along. During a lunch break and ten minutes away from a daunting final physics exam in high school, I sat dejected. A friend asked why, and I explained that I very much doubted that the sum of my knowledge of the subject equalled that required for a passing mark. My friend said a very interesting and useful thing and offered a suggestion. He pointed out that I knew *some* physics. He suggested (in retrospect, a pretty standard suggestion) that I should go through the exam quickly from front to back and answer every question to which I knew the answer or understood the solution. This is a great strategy if you think that you will be time-constrained because it allows the best use of time and does not leave questions unanswered that you could have gotten right. It lets the person who knows the subject reasonably well, but who will have a hard time finishing in the allotted time, get a better result. But then I explained that time was not my problem; I just did not *know* enough. He said words to this effect: 'You're not doing it just to save time; you're doing it to learn physics.' His idea was that when you exercised the knowledge that you had, you filled gaps by connecting ideas and gaining larger insights. After your first run through the easy questions, you would have learned enough to answer a few more. Answering one complicated question could give you enough ideas to answer maybe even a few fairly simple ones to which you did not know the answers going in. I passed. That afternoon, I learned a little physics and I learned important things about solving problems and addressing challenges. Every person who reads this book will have learned that same lesson, one way or another. That was just my day for the parting of the veil. Where policy serves slowly, markets are denied value cues. Technologies abound but confuse, and the problem seems intractable. We must perhaps focus on what *can* be done—and *do* it.

Yes, we must hope for the evolution of international agreements that have meaning and relevance. But it is also true that those needs—accepting that they are frustratingly urgent—must play out over the longer term no matter what. The co-existence of those conflicting truths together—urgency and long time frames— seems to paralyze us. We need catalysis. We need politicians and company executives to see how success happens. That seems to be the only way they may come to understand how their support helps and how action gains success and advantage.

Bending the line that relates energy to GDP helps to relieve pressure; that is the story of efficiency. Applying motivation through sanctions (if policies could ever be agreed) would allow some excess capacity in reducing emissions to find those who could most benefit. But it is the wholesale disconnection of energy from emissions that needs to be accomplished, and it appears that it will not be accomplished broadly until we build smaller alliances of actors who are willing to demonstrate how it is done: someone must stand up, walk across the room and ask another to dance. It almost doesn't matter who. These actions may result in the kind of learning that allows a global society to 'pass its physics exam'.

The caution here, as we have discussed at length, is that absolutely the most important thing is to ensure that each and every individual or organization or municipal or national government, or anyone that might engage on the challenge, understands the criteria of sustainable human enterprise in the same way. They do not need to set identical targets. But they need to see sustainability factors identically: 'carbon reduction' must have the same meaning for everyone even if specific carbon goals might vary, for various reasons. 'Social justice' here must have the same meaning as it has there.

Building understanding, agreement, and action are the keys. Those who take actions are saviors of themselves and the rest of us. They save our economy, our environment and our society, all simultaneously. That is because they make it clear that we are not in a trap. If we progress to higher levels of development with inadequate sources of new energy, energy prices will increase and our economy will struggle. But environmental damage will have a similar effect. To the extent that we fail, rising environmental and economic costs reinforce one another to our detriment. To the extent that we succeed, prosperity, social justice, and healthy environments become more easily associated benefits rather than daunting challenges. Failure to achieve the goal of severing the connection between harmful emissions and energy is to *pay* to have a problem that we do not want. We will be buying higher costs in energy and many other things, unnecessary and undesired. And the solutions are only tricky when we think of solving the whole thing at once and of a piece. Finding a way to solve a bit of it at a time is not so hard.

I was sitting in the audience when a Canadian politician named Bob Rae once said that his wife found him to be a person for whom the opposite of talking was waiting. I am guilty of the same thing. But when we think about how global society has 'addressed' the job of bringing the quality of sustainability to the activities in which we engage, is it not true that we are all like that? The discussion has become a cacophony of conflicting views *even where interests are shared.* Governments fight with other governments, who fight with corporations, who fight with interest groups and, in most cases, all to no purpose. If each one of us were to stop talking *at* everyone and talk *with* someone, we would make rapid progress. For example, everywhere in the world there are municipalities, and groups of municipalities, for whom an airport is a vital source of wealth and advantage. The airlines that serve that airport are their natural allies. There is almost certainly a local resource that could be exploited in a sustainable way to produce fuel. There is also certainly a company somewhere in the world that knows how to accomplish that. But people will ask if that should be the priority. Should we invest our effort that way? Would someone else gain an advantage for which they had not paid us? Should we do it if others do not bother? Should we do it now if there is someone else who should be helping? We ask ourselves these same questions over and over again in all sorts of circumstances. But if we know that the larger job of making all of human society sustainable has to get done, do we really care about all of the reasons why we might not want to get started? They exist. Some of them seem valid. But are they relevant?

I spent my whole working life in aviation and remain engaged and interested in its future. I am convinced that commercial flight can offer great service to the world—service out of all proportion to its cost. I saw the benefits every working day. But humans and their society live on energy. Exploitation of energy resource is a pretty apposite proxy for wealth and advantage. Aviation offers so much because it is right at the front in terms of services that use energy. It is one of the happiest and most productive uses of energy. But it is also one of the most exposed, when we consider the likely future costs of energy that cannot be sourced sustainably. Commercial flight is emblematic of how our society has created its vulnerabilities.

We cannot assume that the industry will continue in its current shape and size—things could go quite wrong; hard realities may dictate a smaller and more expensive service. Airlines and the societies that they serve should undertake to understand their economic interests in a complete way. When governments agree that aviation needs new sustainable sources of flight energy but grumble that it would cost money to help them achieve that, the only answer is 'Yes, it would'. But that is only part of the answer; the other part is, 'But there is an enormous payoff out of all proportion to the effort and expense.'

We have decades ahead of us in completing our tasks, but getting started is a different thing. No matter whether we make the case on selfish interest or moral obligation, there is great urgency. This industry and the societies that depend upon it must achieve real progress in embracing a comprehensive conception of sustainability and applying that to securing alternative sources of flight energy. If part of the answer is getting help, we need to identify that needed assistance and secure it. The number of years or even months before that must begin to happen is zero. There is no time left. None.

Bibliography

Adams, W.M. 2009. *Green Development: Environment and Sustainability in a Developing World*. 3rd Edition. London, U.K. and New York, NY: Routledge.

Advisory Council for Aeronautics Research in Europe. 2010. *Clean Sky: First Interim Evaluation Panel Report*. s.l.: ACARE. Available at: http://www.cleansky.eu/sites/default/files/documents/csju-first-interim-evaluation-20101215.pdf [accessed: March 11, 2014].

Advisory Council for Aviation Research and Innovation in Europe. 2012. *Strategic Research and Innovation Agenda, Volume 1: Realising Europe's Vision for Aviation*. s.l.: ACARE. Available at: http://www.acare4europe.com/sites/acare4europe.org/files/attachment/SRIA%20Volume%201.pdf [accessed: March 11, 2014].

Aerospace Industries Association. 2013. *2013 Year-End Review and Forecast*. Arlington, VA: AIA. Available at: http://www.aia-aerospace.org/assets/2013_AIA_Annual_report_webversion.pdf [accessed: April 17, 2014].

Air Transport Action Group. 2009. *Beginner's Guide to Aviation Biofuels*. Geneva, Switzerland: ATAG. Available at: http://www.enviro.aero/content/upload/file/beginnersguide_biofuels_webres.pdf [accessed: April 8, 2014].

Airbus SAS. 2011. *Delivering the Future: Airbus Global Market Forecast 2011–2030 (full version)*. Blagnac, France: Airbus SAS. Available at: http://www.eads.com/dms/eads/int/en/investor-relations/documents/2011/Presentations/2011-2030_Airbus_full_book_delivering_the_future.pdf [accessed: March 24, 2014].

Airbus SAS. 2012. *Navigating the Future: Global Market Forecast 2012–2031*. Blagnac, France: Airbus SAS.

Airbus SAS. 2014. *Airbus Aircraft 2014 Average List Prices (mio USD)*. [Online]. Airbus. Available at: http://www.airbus.com/presscentre/corporate-information/key-documents/?eID=dam_frontend_push&docID=36716 [accessed: March 24, 2014].

Alderman & Company. 2011. Oil price volatility and aircraft orders. *Air Transport News*. [Online, May 10, 2011]. Available at: http://www.atn.aero/analysis.pl?id=1074&keys=alderman (original post); http://www.aviainform.org/industrynews/14-industrynews/1536-oil-price-volatility-and-aircraft-orders.html (repost, more detail) [accessed: March 11, 2014].

Anseeuw, L. et al. 2012. *Land Rights and The Rush for Land: Findings of the Global Commercial Pressures on Land Research Project*. Rome, Italy: International Land Coalition. Available at: http://www.landcoalition.org/sites/default/files/publication/1205/ILC%20GSR%20report_ENG.pdf [accessed: April 8, 2014].

Archer, D. 2005. Fate of fossil fuel CO_2 in geologic time. *Journal of Geophysical Research–Oceans (Online)*, 110, C09S05 (6 pages). Available at: http://dx.doi. org/10.1029/2004JC002625 (subscription required) [accessed: March 11, 2014].

The Australian. 2008. Virgin biofuel flight "a stunt". *The Australian*. [Online, February 26, 2008]. Available at: http://www.theaustralian.com.au/business/ aviation/virgin-biofuel-flight-a-stunt/story-e6frg95x-1111115643759 [accessed: April 23, 2014].

Ban, K. 2010. *Remarks of the Secretary-General to Media on the First Meeting of the Global Sustainability Panel*. [Online, September 19, 2010]. United Nations. Global Sustainability Panel. Available at: http://www.un.org/wcm/content/site/ climatechange/lang/en/pages/gsp/media/gsp_first_meeting [accessed: March 11, 2014].

Barbose, G. et al. 2013. *Tracking the Sun VI: An Historical Summary of the Installed Price of Photovoltaics in the United States from 1998 to 2012*. Berkeley, CA.: Lawrence Berkeley National Laboratory and United States Department of Energy. Available at: http://emp.lbl.gov/sites/all/files/lbnl-6350e.pdf [accessed: April 18, 2014].

Barrow, B. 2006. Flying on holiday "a sin" says Bishop. *Mail Online [Daily Mail]*. [Online, July 23, 2006]. Available at: http://www.dailymail.co.uk/news/ article-397228/Flying-holiday-sin-says-bishop.html [accessed: March 11, 2014].

Bastos Lima, M.G. and Gupta, J. 2013. The policy context of biofuels: a case of non-governance at the global level? *Global Environmental Politics*, 13 (2), 46–64. Available at: http://dx.doi.org/doi:10.1162/GLEP_a_00166 (subscription required) [accessed: March 11, 2014].

Beddoe, R. et al. 2009. Overcoming systemic roadblocks to sustainability: the evolutionary redesign of worldviews, institutions, and technologies. *Proceedings of the National Academy of Sciences of the USA*, 106 (8), 2483–9. Available at: http://www.pnas.org/content/106/8/2483.long [accessed: March 11, 2014].

Boeing Company. 2004. *Fuel Conservation. Flight Operations Engineering, Boeing Commercial Airplanes*. Seattle, WA: Boeing. Available at: http://www. smartcockpit.com/download.php?path=docs/&file=Fuel_Conservation.pdf [accessed: March 11, 2014].

Boeing Company. 2013. *Current Market Outlook 2013–2032*. Seattle, WA: Boeing Commercial Airplanes. Available at: http://www.boeing.com/assets/pdf/ commercial/cmo/pdf/Boeing_Current_Market_Outlook_2013.pdf [accessed: March 11, 2014].

Boeing Company. 2014a. *737 Family: About the 737 Family*. [Online]. Boeing. Available at: http://www.boeing.com/commercial/737family/background.html [accessed: March 24, 2014].

Boeing Company. 2014b. *747 Family: About the 747 Family*. [Online]. Boeing. Available at: http://www.boeing.com/commercial/747family/background.html [accessed: March 24, 2014].

Bowen, A. et al. 2012. Why do economists describe climate change as a "market failure"? *theguardian.com*. [Online, May 21, 2012]. Available at: http://www.theguardian.com/environment/2012/may/21/economists-climate-change-market-failure [accessed: April 3, 2014].

Boyne, W.J. 2006. The converging paths of Whittle and von Ohain. *Airforce Magazine*, 89 (1), 70–4. Available at: http://www.airforce-magazine.com/MagazineArchive/Documents/2006/January%202006/0106engines.pdf [accessed: March 11, 2014].

Budd, L., Griggs, S. and Howarth, D. 2013. *Sustainable Aviation Futures*. Bingley, U.K.: Emerald Group Publishing.

Canada. Foreign Affairs and International Trade Canada. 2013. *Aerospace Industries: Canada's Competitive Advantages*. Ottawa, ON: Government of Canada. Available at: http://www.international.gc.ca/investors-investisseurs/assets/pdfs/download/canada_aerospace_2013_WCAG.pdf [accessed: April 17, 2014].

CAPA–Centre for Aviation. 2012. *More Subdued Outlook for China in Short Term but Some Mighty Changes Afoot*. [Online, April 18, 2012]. CAPA. Available at: http://centreforaviation.com/analysis/more-subdued-outlook-for-china-in-short-term-but-some-mighty-changes-afoot-71891 [accessed: March 20, 2014].

Carrington, D. 2013. Europe "stigmatising" Canada by labelling tar sands oil highly polluting. *theguardian.com*. [Online, November 19, 2013]. Available at: http://www.theguardian.com/environment/2013/nov/19/canada-tar-sands-oil-eu [accessed: March 11, 2014].

Carson, R. 1962. *Silent Spring*. Boston, MA: Houghton-Mifflin.

Central Intelligence Agency (US). continuously updated. *World Factbook*. [Online]. CIA. Available at: https://www.cia.gov/library/publications/the-world-factbook/index.html [accessed: May 26, 2014].

Christoff, P. 1996. Ecological modernisation, ecological modernities. *Environmental Politics*, 5(3), 476–500. Available at: http://dx.doi.org/10.1080/09644019608414283 (subscription required) [accessed: March 11, 2014].

Commercial Aviation Alternative Fuels Initiative. 2013. *CAAFI Alternative Jet Fuel Environmental Sustainability Overview*. (version 4.0). s.l.: CAAFI. Available at: http://www.caafi.org/information/pdf/Sustainability_Guidance__Posted_2013_07.pdf [accessed: April 17, 2014].

Commission of the European Communities. 2007. *Communication From the Commission to the Council and the European Parliament. Renewable Energy Road Map: Renewable Energies in the 21st Century: Building a More Sustainable Future*. (COM(2006) 848 final). Brussels, Belgium: EC. Available at: http://eur-lex.europa.eu/LexUriServ/LexUriServ.do?uri=COM:2006:0848:FIN:EN:PDF [accessed: March 11, 2014].

Committee on Climate Change. 2009. *Meeting the UK Aviation Target—Options for Reducing Emissions to 2050*. London, U.K.: CCC. Available at: http://www.theccc.org.uk/publication/meeting-the-uk-aviation-target-options-for-reducing-emissions-to-2050/ [accessed: March 11, 2014].

Committee on Climate Change. 2013. *Factsheet: Aviation.* [Online, April 2013]. CCC. Available at: http://www.theccc.org.uk/wp-content/uploads/2013/04/Aviation-factsheet.pdf [accessed: May 4, 2014].

Conference Board of Canada. 2013. *Greenhouse Gas (GHG) Emissions.* [Online, January 2013]. Conference Board. Available at: http://www.conferenceboard.ca/hcp/details/environment/greenhouse-gas-emissions.aspx [accessed: August 15, 2014].

Conrado, R.J. et al. 2013. Electrofuels: a new paradigm for renewable fuels, in *Advanced Biofuels and Bioproducts*, edited by J.W. Lee. New York, NY: Springer Science+Business Media, 1037–64. Available at: http://dx.doi.org/10.1007/978-1-4614-3348-4_38 (subscription required) [accessed: March 24, 2014].

Costanza, R. et al. 1997. The value of the world's ecosystem services and natural capital. *Nature*, 387 (6630), 253–60. Available at: http://www.nature.com/nature/journal/v387/n6630/pdf/387253a0.pdf [accessed: April 3, 2014].

Council on Sustainable Biomass Production. 2012. *A Comprehensive Standard and National Certification Program for Sustainable Production of Cellulosic Biomass and Bioenergy.* (NRCS Agreement: #69-3A75-10-178). Washington, DC: Meridian Institute. Available at: http://web.ornl.gov/sci/ees/cbes/News/Final%20CSBP%20Standard%2020120612.pdf [accessed: April 8, 2014].

Coward, H. and Weaver, A.J. 2004. *Hard Choices: Climate Change in Canada.* Waterloo, ON: Centre for Studies in Religion and Society / WLU Press.

Cramer, J. et al. 2006. *Criteria for Sustainable Biomass Production: Final Report of the Project Group "Sustainable Production of Biomass".* s.l.: Energy Transition's Interdepartmental Programme Management [Netherlands government]. Available at: http://www.globalproblems-globalsolutions-files.org/unf_website/PDF/criteria_sustainable_biomass_prod.pdf [accessed: April 8, 2014].

Cramer, J. et al. 2007. *Testing Framework for Sustainable Biomass: Final Report from the Project Group "Sustainable Production of Biomass".* s.l.: Energy Transition's Interdepartmental Programme Management [Netherlands government]. Available at: http://iet.jrc.ec.europa.eu/remea/sites/remea/files/testing_framework_biomass.pdf [accessed: April 8, 2014].

Diederiks-Verschoor, I.H.P. 2006. *Introduction to Air Law.* 8th Edition. Alphen aan den Rijn, Netherlands: Kluwer Law International.

Diringer, E. et al. 2009. *Summary: Copenhagen Climate Summit. Fifteenth Session of the Conference of the Parties to the United Nations Framework Convention on Climate Change and Fifth Session of the Meeting of the Parties to the Kyoto Protocol; December 7–18, 2009. Copenhagen, Denmark.* Arlington, VA: C2ES–Center for Climate and Energy Solutions. Available at: http://www.c2es.org/international/negotiations/cop-15/summary [accessed: April 5, 2014].

Doyle, Sir A.C. 1890. *The Sign of Four.* London, U.K.: Spencer Blackett.

Dyer, G. 2008. *Climate Wars.* Toronto, ON: Random House Canada.

Ehrlich, P.R. and Holdren, J.P. 1972. Critique: One-Dimensional Ecology. *Bulletin of the Atomic Scientists*, 28 (5), 16, 18–27. Available at: http://books.google.ca/books?id=pwsAAAAAMBAJ&printsec=frontcover&source=gbs_ge_summary_r&cad=0#v=onepage&q&f=false [accessed: April 15, 2014].

Einstein, A. 1920. *Relativity: the Special and the General Theory; a Popular Exposition*, translated by R. W. Lawson. New York, NY: Henry Holt.

Ellstrand, N. 2003. *Dangerous Liaisons?: When Cultivated Plants Mate With Their Wild Relatives*. Baltimore, MD: Johns Hopkins University Press.

Espagne, E. et al. 2012. *Disentangling the Stern/Nordhaus Controversy: Beyond the Discounting Clash (Nota di Lavoro 61.2012)*. Milan, Italy: Fondazione Eni Enrico Mattei. (Climate Change and Sustainable Development Series). Available at: http://www.feem.it/userfiles/attach/20129121056434NDL2012-061.pdf [accessed: April 3, 2014].

European Commission. 2011. *Commission Implementing Decision of 19 July 2011: On the Recognition of the "Roundtable of Sustainable Biofuels EU RED" Scheme for Demonstrating Compliance with the Sustainability Criteria under Directives 2009/28/EC and 2009/30/EC of the European Parliament and of the Council*. (2011/435/EU). Brussels, Belgium: European Union. Available at: http://eur-lex.europa.eu/LexUriServ/LexUriServ.do?uri=OJ:L:2011:190:0073:0074:EN:PDF [accessed: April 9, 2014].

European Parliament and Council. 1998. *Directive 98/70/EC of the European Parliament and of the Council of 13 October 1998 Relating to the Quality of Petrol and Diesel Fuels and Amending Council Directive 93/12/eec*. (Directive 98/70/EC). Brussels, Belgium: European Union. Available at: http://eur-lex.europa.eu/legal-content/EN/TXT/HTML/?uri=CELEX:31998L0070&from=EN [accessed: April 9, 2014].

European Parliament and Council. 2009a. *Directive 2009/28/EC of the European Parliament and of the Council of 23 April 2009, on the Promotion of the Use of Energy from Renewable Sources and Amending and Subsequently Repealing Directives 2001/77/EC and 2003/30/EC*. (Directive 2009/28/EC). Brussels, Belgium: European Union. Available at: http://eur-lex.europa.eu/LexUriServ/LexUriServ.do?uri=OJ:L:2009:140:0016:0062:EN:PDF [accessed: April 8, 2014].

European Parliament and Council. 2009b. *Directive 2009/30/EC of the European Parliament and of the Council of 23 April 2009 Amending Directive 98/70/EC as Regards the Specification of Petrol, Diesel and Gas–Oil and Introducing a Mechanism to Monitor and Reduce Greenhouse Gas Emissions and Amending Council Directive 1999/32/EC as Regards the Specification of Fuel Used by Inland Waterway Vessels and Repealing Directive 93/12/EEC*. (Directive 2009/30/EC). Brussels, Belgium: European Union. Available at: http://eur-lex.europa.eu/LexUriServ/LexUriServ.do?uri=OJ:L:2009:140:0088:0113:EN:PDF [accessed: April 9, 2014].

European Union. 2012. *Single European Sky: 10 Years On and Still Not Delivering.* [Online, October 11, 2012]. EU. Available at: http://europa.eu/rapid/press-release_IP-12-1089_en.htm [accessed: March 20, 2014].

Freer, D. 1986a. An aborted take-off for internationalism—1903 to 1919. Special series, part 2. *ICAO Bulletin*, 41 (4), 23–6. Available at: http://www.icao.int/publications/Pages/ICAO-Journal.aspx [accessed: March 28, 2014].

Freer, D. 1986b. Chicago Conference (1944) —Despite uncertainty, the spirit of internationalism soars. Special series, part 7. *ICAO Bulletin*, 41 (9), 42–4. Available at: http://www.icao.int/publications/Pages/ICAO-Journal.aspx [accessed: March 28, 2014].

Freer, D. 1986c. Chicago Conference (1944) —U.K.–U.S. policy split revealed. Special series, part 6. *ICAO Bulletin*, 41 (8), 22–4. Available at: http://www.icao.int/publications/Pages/ICAO-Journal.aspx [accessed: March 28, 2014].

Freer, D. 1986d. A Convention is signed and ICAN is born—1919 to 1926. Special series, part 3. *ICAO Bulletin*, 41 (5), 44–6. Available at: http://www.icao.int/publications/Pages/ICAO-Journal.aspx [accessed: March 28, 2014].

Freer, D. 1986e. En-route to Chicago—1943 to 1944. Special series, part 5. *ICAO Bulletin*, 41 (7), 39–41. Available at: http://www.icao.int/publications/Pages/ICAO-Journal.aspx [accessed: March 28, 2014].

Freer, D. 1986f. Gear up! 1947 to 1957. Special series, part 9. *ICAO Bulletin*, 41 (11), 52–4. Available at: http://www.icao.int/publications/Pages/ICAO-Journal.aspx [accessed: March 28, 2014].

Freer, D. 1986g. Maturity brings new challenges 1957 to 1976. Special series, part 10. *ICAO Bulletin*, 41 (12), 24–6. Available at: http://www.icao.int/publications/Pages/ICAO-Journal.aspx [accessed: March 28, 2014].

Freer, D. 1986h. The PICAO years 1945–1947. Special series, part 8. *ICAO Bulletin*, 41 (10), 36–9. Available at: http://www.icao.int/publications/Pages/ICAO-Journal.aspx [accessed: March 28, 2014].

Freer, D. 1986i. Regionalism is asserted. ICAN's global prospects fade—1926 to 1943. Special series, part 4. *ICAO Bulletin*, 41 (6), 66–8. Available at: http://www.icao.int/publications/Pages/ICAO-Journal.aspx [accessed: March 28, 2014].

Freer, D. 1986j. The roots of internationalism—1783 to 1903. Special series, part 1. *ICAO Bulletin*, 41 (3), 30–2. Available at: http://www.icao.int/publications/Pages/ICAO-Journal.aspx [accessed: March 28, 2014].

Freer, D. 1987a. 203 years in retrospect, 1783 to 1986. Special series, part 12. *ICAO Bulletin*, 42 (2), 23–5. Available at: http://www.icao.int/publications/Pages/ICAO-Journal.aspx [accessed: March 28, 2014].

Freer, D. 1987b. New problems arise, old ones return. Special series, part 11. *ICAO Bulletin*, 42 (1), 32–5. Available at: http://www.icao.int/publications/Pages/ICAO-Journal.aspx [accessed: March 28, 2014].

Gardner, S.M. 2006. A Perfect moral storm: climate change, intergenerational ethics, and the problem of moral corruption. *Environmental Values*, 15 (3), 397–413. Available at: http://www.hettingern.people.cofc.edu/Environmental_

Philosophy_Sp_09/Gardner_Perfect_Moral_Storm.pdf [accessed: April 3, 2014].

Gore, A. 2009. *Our Choice: a Plan to Solve the Climate Crisis*. Emmaus, PA: Rodale.

Government Accountability Office (US). 2009. *Aviation and Climate Change*. Washington, DC: US GAO. Available at: http://www.gao.gov/assets/300/290594.pdf [accessed: March 17, 2014].

Green, J.E. 2009. The potential for reducing the impact of aviation on climate. *Technology Analysis and Strategic Management*, 21 (1), 39–59. Available at: http://dx.doi.org/10.1080/09537320802557269 (subscription required) [accessed: March 20, 2014].

GreenAir Online.com. 2013. *Proposal by BRIC States to Set Up a New Group on a Global MBM for Aviation Approved by ICAO Council*. [Online, December 20, 2013]. GreenAir Communications. Available at: http://www.greenaironline.com/news.php?viewStory=1804 [accessed: April 13, 2014].

Grierson, J. 1971. Britain's first jet aeroplane. *Flight International*, (May 13, 1971), 677–9. Available at: http://www.flightglobal.com/pdfarchive/view/1971/1971%20-%200766.html [accessed: March 23, 2014].

Gunston, B. 2006. *The Development of Jet and Turbine Aero Engines*. 4th Edition. Sparkford, U.K.: Patrick Stephens.

Gurria, A. 2013. *The Climate Challenge: Achieving Zero Emissions*. [Online, October 9, 2013]. Organisation for Economic Co-operation and Development. Available at: http://www.oecd.org/about/secretary-general/the-climate-challenge-achieving-zero-emissions.htm [accessed: April 25, 2014].

Havel, B.F. and Sanchez, G.S. 2012. Toward a global aviation emissions agreement. *Harvard Environmental Law Review*, 36 (2), 351–85. Available at: http://papers.ssrn.com/sol3/papers.cfm?abstract_id=1911508 (prepublication version) [accessed: March 29, 2014].

Heimlich, J.P. 2009. Alternative fuels: why do we need them? in *ICAO Workshop [on] Aviation and Alternative Fuels; Montreal QC, 10–12 February, 2009*. Montreal, QC: International Civil Aviation Organization, n.p.

Heinrich Böll Foundation. 2003. 1992: The Rio earth summit. *World Summit 2002*. [Online, July 18, 2003]. Available at: http://www.worldsummit2002.org/index.htm?http://www.worldsummit2002.org/guide/unced.htm [accessed: March 17, 2014].

Hertogen, A. 2012. Sovereignty as decisional independence over domestic affairs: the dispute over aviation in the EU Emissions Trading System. *Transnational Environmental Law*, 1 (2), 281–301. Available at: http://dx.doi.org/10.1017/S204710251200012X (subscription required) [accessed: April 13, 2014].

Hope, M. 2014. The pitfalls of analysing media coverage of climate change, in three graphs. *The Carbon Brief*. [Online, January 7, 2014]. Available at: http://www.carbonbrief.org/blog/2014/01/the-pitfalls-of-analysing-media-coverage-of-climate-change,-in-three-graphs/ [accessed: April 25, 2014].

Horng, T.-C. 2006. *A Comparative Analysis of Supply Chain Management Practices by Boeing and Airbus: Long-term Strategic Implications.* (M.Sc. Thesis). Cambridge, MA: Massachusetts Institute of Technology. Available at: http://hdl.handle.net/1721.1/38579 (subscription required) [accessed: March 23, 2014].

Hu, J., Yu, F. and Lu, Y. 2012. Application of Fischer-Tropsch synthesis in biomass to liquid conversion. *Catalysts,* 2, 303–26. Available at: http://dx.doi.org/10.3390/catal2020303 [accessed: March 24, 2014].

An Inconvenient Truth: A Global Warming. (film). 2006. Directed by A. Gore. Hollywood, CA: Lawrence Bender Productions and Participant Productions.

Intergovernmental Panel on Climate Change. 2006. *2006 IPCC Guidelines for National Greenhouse Gas Inventories, Volume 2: Energy,* edited by H.S. Eggleston et al. Hayama, Japan: Institute for Global and Environmental Strategies. Available at: http://www.ipcc-nggip.iges.or.jp/public/2006gl/vol2.html [accessed: March 28, 2014].

Intergovernmental Panel on Climate Change. 2007a. *Climate Change 2007: Synthesis Report. An Assessment of the Intergovernmental Panel on Climate Change,* edited by Core Writing Team, R.K. Pachauri and A. Reisinger. Geneva, Switzerland: IPCC. Available at: http://www.ipcc.ch/publications_and_data/ar4/syr/en/mains3.html or http://www.ipcc.ch/pdf/assessment-report/ar4/syr/ar4_syr.pdf [accessed: March 17, 2014].

Intergovernmental Panel on Climate Change. 2007b. *IPCC Fourth Assessment Report: Climate Change 2007 (AR4).* Geneva, Switzerland: IPCC. Available at: http://www.ipcc.ch/publications_and_data/publications_and_data_reports.shtml#1 (links for all AR4 reports) [accessed: March 17, 2014].

Intergovernmental Panel on Climate Change. 2014. *Climate Change 2014: Mitigation of Climate Change; IPCC Working Group III Contribution to AR5 [final draft],* edited by O. Edenhofer et al. Cambridge, U.K. and New York, NY: Cambridge University Press. Available at: http://mitigation2014.org/report/final-draft/ [accessed: May 3, 2014].

International Air Transport Association. 2013. *Resolution on the Implementation of the Aviation "CNG2020" Strategy. [IATA Annual General Meeting, 69th; June 2–4 2013, Cape Town, S.A.].* [Online, June 2013]. IATA. Available at: http://www.iata.org/pressroom/pr/Documents/agm69-resolution-cng2020.pdf [accessed: April 11, 2014].

International Air Transport Association. 2014. *IATA Fact Sheet: Industry Statistics [June 2014].* [Online, March 2014]. IATA. Available at: http://www.iata.org/pressroom/facts_figures/fact_sheets/Documents/industry-facts.pdf [accessed: August 15, 2014].

International Air Transport Association. n.d. *IATA Sustainable Alternative Aviation Fuels Strategy.* [Online]. IATA. Available at: http://www.iata.org/whatwedo/environment/Documents/sustainable-alternative-aviation-fuels-strategy.pdf [accessed: April 17, 2014].

International Civil Aviation Organization. 1944a. *Convention on International Civil Aviation (Chicago Convention)*. Montreal, QC: ICAO. Available at: http://www.icao.int/publications/Documents/7300_1ed.pdf [accessed: March 28, 2014].

International Civil Aviation Organization. 1944b. *International Air Services Transit Agreement, Signed at Chicago, on 7 December 1944 (Transit Agreement)*. Chicago, IL: ICAO. Available at: http://www.mcgill.ca/files/iasl/chicago1944b.pdf [accessed: March 28, 2014].

International Civil Aviation Organization. 1944c. *International Air Transport Agreement Signed at Chicago on 7 December 1944 (Transport Agreement)*. Chicago, IL: ICAO. Available at: http://library.arcticportal.org/1584/1/international_air_transport_agreement_chicago1944c.pdf [accessed: March 28, 2014].

International Civil Aviation Organization. 2010. Resolution A37-19: Consolidated statement of continuing ICAO policies and practices related to environmental protection—climate change, in *International Civil Aviation Organization. Assembly, 37th Session; Montreal, 28 September–8 October 2010*. Montreal, QC: ICAO, I 67–I 74. Available at: http://www.icao.int/publications/Documents/9958_en.pdf [accessed: March 20, 2014].

International Civil Aviation Organization. 2012a. *Flightpath to a Sustainable Future: the ICAO Rio+20 Global Initiative*. [Online, June 18, 2012]. ICAO. Available at: http://www.icao.int/Newsroom/Pages/Flightpath-to-a-sustainable-future-the-ICAO-Rio+20-global-initiative.aspx [accessed: March 24, 2014].

International Civil Aviation Organization. 2012b. *Global Aviation and Our Sustainable Future: International Civil Aviation Organization Briefing for RIO+20*. Montreal, QC: ICAO. Available at: http://www.icao.int/environmental-protection/Documents/RIO+20_booklet.pdf [accessed: August 14, 2014].

International Civil Aviation Organization. 2013a. *Annual Report of the Council, 2012: Documentation for the Session of the Assembly in 2013*. Montreal, QC: ICAO. Available at: http://www.icao.int/publications/Documents/10001_en.pdf [accessed: March 23, 2014].

International Civil Aviation Organization. 2013b. *Dramatic MBM Agreement and Solid Global Plan Endorsements Help Deliver Landmark ICAO 38th Assembly (press release)*. [Online, October 4, 2013]. ICAO. Available at: http://www.icao.int/Newsroom/Pages/mbm-agreement-solid-global-plan-endoresements.aspx [accessed: April 13, 2014].

International Civil Aviation Organization. 2013c. *ICAO Predicts Continued Traffic Growth Through 2015 (News Release)*. [Online, July 2013]. ICAO. Available at: http://www.icao.int/Newsroom/News%20Doc%202013/COM.24.13.EN.pdf [accessed: March 17, 2014].

International Civil Aviation Organization. 2013d. *Report of the Executive Committee on Agenda Item 17 (Section on Climate Change). International Civil Aviation Organization Assembly, 38th Session; Montreal, 24 September–4 October 2013*. Montreal, QC: ICAO. Available at: http://www.icao.int/Meetings/a38/Documents/WP/wp430_en.pdf [accessed: August 15, 2014].

International Civil Aviation Organization. 2013e. *Resolutions Adopted By the Assembly (Provisional Edition). ICAO Assembly, 38th Session; Montreal, 24 September–4 October, 2013.* Montreal, QC: ICAO. Available at: http://www.icao.int/Meetings/a38/Documents/Resolutions/a38_res_prov_en.pdf [accessed: August 15, 2014].

International Civil Aviation Organization Executive Committee. 2013. *Addressing CO_2 Emissions from Aviation, Revision no. 2. Working Paper, Agenda Item 17: Environmental Protection. International Civil Aviation Organization Assembly, 38th Session; Montreal, 24 September–4 October 2013.* Montreal, QC: ICAO. Available at: http://www.icao.int/Meetings/a38/Documents/WP/wp068_rev2_en.pdf [accessed: August 15, 2014].

International Consortium of Investigative Journalists. 2009. *Interactive: Global Greenhouse Gas Emissions by Country.* [Online, December 7, 2009]. ICIJ. Available at: http://www.icij.org/project/global-climate-change-lobby/interactive-global-greenhouse-gas-emissions-country [accessed: April 5, 2014].

International Coordinating Council of Aerospace Industries Associations. 2009. *Continuous Improvement in Aircraft Fuel Efficiency. Working Paper, Agenda Item 1: Environmental Sustainability and Interdependencies. ICAO Conference on Aviation and Alternative Fuels; Rio de Janeiro, Brazil, 16 to 18 November, 2009.* Montreal, QC: International Civil Aviation Organization. Available at: http://www.icao.int/Meetings/caaf2009/Documents/CAAF-09_WP008_en.pdf [accessed: August 13, 2014].

International Energy Agency. 2012. *Energy Technology Perspectives 2012: Pathways to a Clean Energy System.* Paris, France: IEA. Available at: http://http://www.iea.org/etp/ (subscription required to access full report) [accessed: May 3, 2014].

International Energy Agency. 2013. *Key World Energy Statistics.* Paris, France: IEA. Available at: http://www.iea.org/publications/freepublications/publication/KeyWorld2013.pdf [accessed: April 25, 2014].

ITN (Independent Television News). 2013. OECD: We must eliminate fossil fuel emissions (online video). *The Telegraph.* [Online, October 9, 2013]. Available at: http://www.telegraph.co.uk/earth/environment/climatechange/10368392/OECD-We-must-eliminate-fossil-fuel-emissions.html [accessed: April 25, 2014].

Joint Inspection Group. 2012. Aviation fuel quality requirements for jointly operated systems (AFQRJOS). *JIG Joint Inspection Group Product Specifications,* 2012 (Issue 26, Bulletin no. 51), 1–6. Available at: http://www.jigonline.com/wp-content/uploads/2012/05/Bulletin-51-AFQRJOS-Issue-26-May-2012.pdf [accessed: March 23, 2014].

Joule Unlimited. 2014. *Why Joule?* [Online]. Joule Unlimited. Available at: http://www.jouleunlimited.com/why-joule [accessed: March 24, 2014].

Kyprianidis, K.G. et al. 2011. Assessment of future aero-engine designs with intercooled and intercooled recuperated cores. *Journal of Engineering for Gas Turbines and Power (Online),* 133 (1), 011701 (10 pages). Available at: http://dx.doi.org/10.1115/1.4001982 (subscription required) [accessed: March 20, 2014].

Lane, J. 2012. Aviation biofuels: which airlines are doing what, with whom? *Biofuels Digest*. [Online, June 5, 2012]. Available at: http://www.biofuelsdigest. com/bdigest/2012/06/05/aviation-biofuels-which-airlines-are-doing-what-with-whom/ [accessed: May 3, 2014].

Lane, J. 2013. Obama messes with the RFS. *Biofuels Digest*. [Online, November 17, 2013]. Available at: http://www.biofuelsdigest.com/bdigest/2013/11/17/obama-messes-with-the-rfs/ [accessed: April 23, 2014].

Lappé, F.M. 1975. *Diet for a Small Planet*. New York, NY: Ballantine.

Le Quéré, C. et al. 2013. The global carbon budget 1959–2011. *Earth System Science Data*, 5 (1), 165–85. Available at: http://www.earth-syst-sci-data. net/5/165/2013/essd-5-165-2013.pdf [accessed: May 1, 2014].

Le Treut, H. et al. 2007. Historical overview of climate change, in *Climate Change 2007: The Physical Science Basis. Contribution of Working Group I to the Fourth Assessment Report of the Intergovernmental Panel on Climate Change*, edited by S. Solomon et al. Cambridge, U.K. and New York, NY: Cambridge University Press, 93–128. Available at: http://www.ipcc.ch/pdf/assessment-report/ar4/wg1/ar4-wg1-chapter1.pdf or http://www.ipcc.ch/publications_ and_data/ar4/wg1/en/ch1.html [accessed: March 17, 2014].

Lee, D.S. et al. 2009. Aviation and global climate change in the 21st century. *Atmospheric Environment*, 43 (22/23), 3520–37. Available at: http://dx.doi. org/10.1016/j.atmosenv.2009.04.024 (subscription required) [accessed: March 17, 2014].

Lee, D.S., Lim, L.L. and Owen, B. 2013a. *Bridging the Aviation CO$_2$ Emissions Gap: Why Emissions Trading is Needed*. Manchester, U.K.: Manchester Metropolitan University. Available at: http://www.cate.mmu.ac.uk/wp-content/ uploads/Bridging_the_aviation_emissions_gap_010313.pdf [accessed: April 25, 2014].

Lee, D.S., Lim, L.L. and Owen, B. 2013b. *Mitigating Future Aviation CO$_2$ Emissions—"Timing is Everything"*. Manchester, U.K.: Manchester Metropolitan University. Available at: http://www.cate.mmu.ac.uk/docs/ mitigating-future-aviation-co2-emissions.pdf [accessed: April 10, 2014].

Leggett, J., Pepper, W.J. and Swart, R.J. 1992. Emissions scenarios for IPCC: an update, in *Climate Change 1992: Supplementary Report to the IPCC Scientific Assessment. Report Prepared for Intergovernmental Panel on Climate Change by Working Group I Combined with Supporting Scientific Material*, edited by J.T. Houghton, B.A. Callender and S.K. Varney. Cambridge, U.K.: Cambridge University Press, 69–95. Available at: http://www.ipcc.ch/publications_ and_data/publications_ipcc_supplementary_report_1992_wg1.shtml#. UPmXMM0sluI [accessed: March 17, 2014].

Liang, L. and James, A.D. 2009. The low-cost carrier model in China: the adoption of a strategic innovation. *Technology Analysis & Strategic Management*, 21 (1), 129–48. Available at: http://dx.doi.org/10.1080/09537320802557384 (subscription required) [accessed: March 20, 2014].

Lissitzyn, O.J. 1964. Bilateral agreements on air transport. *Journal of Air Law and Commerce*, 30 (3), 248–63. Available at: http://www.heinonline.org/HOL/Page?handle=hein.journals/jalc30&collection=journals&index=journals/jalc&id=258 (subscription required) [accessed: March 28, 2014].

Luke, T.W. 2005. Neither sustainable nor development: reconsidering sustainability in development. *Sustainable Development*, 13 (4), 228–38. Available at: http://dx.doi.org/10.1002/sd.284 (subscription required) [accessed: March 30, 2014].

MacNeill, J. 2010. Introductions: 1, in *Cents and Sustainability: Securing Our Common Future by Decoupling Economic Growth from Environmental Pressures*, edited by M.H. Smith, K. Hargroves and C. Desha. London, U.K. and Washington, DC: Earthscan, xxxiii–vii.

McKibbin, W.J. and Stegman, A. 2005. *Convergence and Per Capita Carbon Emissions*. Sydney, Australia: Lowy Institute. Available at: http://www.lowyinstitute.org/files/pubfiles/Mckibbin%2C_Convergence_and_per_capita.pdf [accessed: April 15, 2014].

Meadows, D.H. 1972. *Limits to Growth: A Report for the Club of Rome's Project on the Predicament of Mankind*. New York, NY: Universe Books.

Mitchell, A. 2009. *Seasick: Ocean Change and the Extinction of Life on Earth*. Chicago, IL: University of Chicago Press.

Myhre, G. et al. 2013. Anthropogenic and natural radiative forcing, in *Climate Change 2013: the Physical Science Basis. Contribution of Working Group I to the Fifth Assessment Report of the Intergovernmental Panel on Climate Change*, edited by T.F. Stocker et al. Cambridge, U.K. and New York, NY: Cambridge University Press, 659–740. Available at: http://www.climatechange2013.org/images/report/WG1AR5_Chapter08_FINAL.pdf [accessed: March 17, 2014].

Nakicenovic, N. and Swart, R.J. 2000. *Special Report on Emissions Scenarios: A Special Report of Working Group III of the Intergovernmental Panel on Climate Change*. Cambridge, U.K.: Cambridge University Press (prepared and published to web by GRID-Arendal, 2001). (IPCC Special Reports on Climate Change). Available at: http://www.ipcc.ch/ipccreports/sres/emission/index.php?idp=0 [accessed: March 17, 2014].

National Aeronautics and Space Administration. 2010. *Beauty of Future Airplanes is More Than Skin Deep*. [Online, May 17, 2010]. NASA. Available at: http://www.nasa.gov/topics/aeronautics/features/future_airplanes.html [accessed: March 20, 2014].

National Research Council (US). 2011. *Renewable Fuel Standard: Potential Economic and Environmental Effects of U.S. Biofuel Policy*. Washington, DC: National Academies Press. Available at: http://www.nap.edu/catalog.php?record_id=13105 [accessed: April 15, 2014].

The Nature of Things. (television series). 1960–current. Hosted by D. Suzuki. s.l.: Canadian Broadcasting Corporation (CBC) and Discovery Channel.

Nelson, M. et al. 2010. Closed ecological systems, space life support and biospherics, in *Environmental Biotechnology*, edited by L.K. Wang et al. New York, NY: Humana Press, 517–65.

Nordhaus, W.D. 2007. A Review of the Stern Review on the Economics of Climate Change. *Journal of Economic Literature*, 45 (3), 686–702. Available at: http://dx.doi.org/10.1257/jel.45.3.686 (subscription required) [accessed: April 3, 2014].

Nordhaus, W.D. 2012. Why the global warming skeptics are wrong. *New York Review of Books [blog]*. [Online, March 22, 2012]. Available at: http://www.nybooks.com/articles/archives/2012/mar/22/why-global-warming-skeptics-are-wrong/ [accessed: April 3, 2014].

Norris, G. 2003. Sonic cruiser is dead—long live super efficient? *Flightglobal*. [Online, January 7, 2003]. Available at: http://www.flightglobal.com/articles/2003/01/07/159915/sonic-cruiser-is-dead-long-live-super-efficient.html [accessed: March 20, 2014].

Norton, B.G. 2005. *Sustainability: a Philosophy of Adaptive Ecosystem Management*. Chicago, IL: University of Chicago Press.

Organisation for Economic Co-operation and Development. 2013. *Climate and Carbon: Aligning Prices and Policies. (OECD Environment Policy Paper no. 1, October 2013)*. Paris, France: OECD. Available at: http://www.oecd-ilibrary.org/environment-and-sustainable-development/climate-and-carbon_5k3z11hjg6r7-en [accessed: April 25, 2014].

Owen, B., Lee, D.S. and Ling, L. 2010. Flying into the future. *Environmental Science and Technology*, 44 (7), 2255–60. Available at: http://pubs.acs.org/doi/pdfplus/10.1021/es902530z [accessed: March 17, 2014].

Oxford English Dictionary. 1971. *Compact Edition of the Oxford English Dictionary*, edited by J.A.H. Murray. Oxford, U.K.: Clarendon Press.

Peace, J. and Weyant, J. 2008. *Insights Not Numbers: The Appropriate Use of Economic Models (white paper)*. Arlington, VA: Pew Center on Global Climate Change (now C2ES–Center for Climate and Energy Solutions). Available at: http://www.c2es.org/docUploads/insights-not-numbers.pdf [accessed: April 11, 2014].

Penner, J.E. et al. 1999. *Aviation and the Global Atmosphere*. Cambridge, U.K.: Cambridge University Press. Available at: http://www.ipcc.ch/ipccreports/sres/aviation/index.php?idp=0 [accessed: May 3, 2014].

Pentland, W. 2013. No end in sight for Spain's escalating solar crisis. *Forbes.com*. [Online, August 16, 2013]. Available at: http://www.forbes.com/sites/williampentland/2013/08/16/no-end-in-sight-for-spains-escalating-solar-crisis/ [accessed: April 10, 2014].

Pope, J., Annandale, D. and Morrison-Saunders, A. 2004. Conceptualising sustainability assessment. *Environmental Impact Assessment Review*, 24 (6), 595–616. Available at: http://dx.doi.org/10.1016/j.eiar.2004.03.001 (subscription required) [accessed: April 6, 2014].

Poynter, J. 2009. Life in Biosphere 2 (online video). *Ted Talks*. [Online, March 2009]. Available at: http://www.ted.com/talks/jane_poynter_life_in_biosphere_2 [accessed: April 3, 2014].

Pratt & Whitney Aircraft Group. 1974. *Aircraft Gas Turbine Engine and Its Operation*. East Hartford, CT: Pratt & Whitney Aircraft Group.

Pullum, G.K. 2005. Yoda's syntax the Tribune analyzes; supply more details I will! *Language Log*. [Online, May 18, 2005]. Available at: http://itre.cis.upenn.edu/~myl/languagelog/archives/002173.html [accessed: March 31, 2014].

Rabello, M.L. and Gamarski, R. 2013. Brazil Aviation Minister says he's pushing for cheaper jet fuel. *Bloomberg.com*. [Online, September 12, 2013]. Available at: http://www.bloomberg.com/news/2013-09-12/brazil-aviation-minister-says-he-s-pushing-for-cheaper-jet-fuel.html [accessed: April 23, 2014].

Redclift, M. 2005. Sustainable development (1987–2005): an oxymoron comes of age. *Sustainable Development*, 13 (4), 212–27. Available at: http://dx.doi.org/10.1002/sd.281 (subscription required) [accessed: March 31, 2014].

Reed, M.E. 2007. Coal and biomass co-conversion to transportation fuels, in *Gasification Technologies Conference: Biomass Gasification; San Francisco, September 15–17, 2007*. Arlington, VA: Gasification Technologies Council, n.p. Available at: http://www.gasification.org//uploads/eventLibrary/52REED.pdf [accessed: August 15, 2014].

Riedel, M., Scott, R. and Junyang, J. 2013. *China—Peoples Republic of: Biofuels Annual*. Washington, DC: USDA Foreign Agricultural Service. Available at: http://gain.fas.usda.gov/Recent%20GAIN%20Publications/Biofuels%20Annual_Beijing_China%20-%20Peoples%20Republic%20of_9-9-2013.pdf [accessed: May 4, 2014].

Robinson, J. 2004. Squaring the circle? Some thoughts on the idea of sustainable development. *Ecological Economics*, 48 (4), 369–84. Available at: http://dx.doi.org/10.1016/j.ecolecon.2003.10.017 (subscription required) [accessed: April 1, 2014].

Roca, M. 2013. Spain hurts solar with plan to penalize power producers. *Bloomberg.com*. [Online, August 1, 2013]. Available at: http://www.bloomberg.com/news/2013-08-01/spain-hurts-solar-with-plan-to-penalize-power-producers.html [accessed: April 10, 2014].

Roundtable on Sustainable Biofuels. 2010. *RSB Principles & Criteria for Sustainable Biofuel Production*. (RSB-STD-01-001 (Version 2.0)). Lausanne, Switzerland: RSB. Available at: http://rsb.org/pdfs/standards/11-03-08-RSB-PCs-Version-2.pdf [accessed: April 9, 2014].

Royal Society of London. 2011. *Solar Nanotech: Putting Sunshine in the Tank: Using Nanotechnology to Make Solar Fuel*. [Online, June 7, 2011]. Royal Society. Available at: http://royalsociety.org/summer-science/2011/solar-nanotech/ [accessed: March 24, 2014].

Rutherford, D. and Zeinali, M. 2009. *Efficiency Trends for New Commercial Jet Aircraft 1960–2008*. Washington, DC: International Council on Clean Transportation. Available at: http://www.theicct.org/sites/default/files/publications/ICCT_Aircraft_Efficiency_final.pdf [accessed: March 24, 2014].

Schnepf, R. and Yacobucci, B.D. 2013. *Renewable Fuel Standard (RFS): Overview and Issues*. Washington, DC: Congressional Research Service. Available at: http://www.fas.org/sgp/crs/misc/R40155.pdf [accessed: April 14, 2014].

Schumann, U. 2010. Recent research results on the climate impact of contrail cirrus and mitigation options, in *En-Route to Sustainability: ICAO Colloquium on Aviation and Climate Change; Montreal QC, May 11–14, 2010*. Montreal, QC: International Civil Aviation Organization, n.p. Available at: http://www.icao.int/Meetings/EnvironmentalColloquium/Documents/2010-Colloquium/1_Schumann_ContrailMitigation.pdf [accessed: August 15, 2014].

Schumann, U., Graf, K. and Mannstein, H. 2011. Potential to reduce the climate impact of aviation by flight level changes, in *AIAA Atmospheric Space Environments Conference, 3rd; Honolulu, June 27–30, 2011*. Reston, VA: American Institute of Aeronautics and Astronautics, AIAA 2011-3376 1–22. Available at: http://www.dlr.de/pa/Portaldata/33/Resources/dokumente/cocip/Schumann_etal_AIAA_2011_3376.pdf [accessed: April 10, 2014].

Searchinger, T. et al. 2008. Use of U.S. croplands for biofuels increases greenhouse gases through emissions from land-use change. *Science*, 319 (5867), 1238–40. Available at: http://www.sciencemag.org/content/319/5867/1238.full.pdf?sid=69215b39-5a89-45eb-880e-9ffb5ff2269d [accessed: April 8, 2014].

Single European Sky ATM Research Consortium. 2009. *European Air Traffic Management Master Plan: Edition 1, 30 March 2009*. Brussels, Belgium: SESAR. Available at: http://www.sesarju.eu/sites/default/files/documents/reports/European_ATM_Master_Plan.pdf [accessed: March 20, 2014].

Single European Sky ATM Research Consortium. 2012. *European ATM Master Plan: The Roadmap for Sustainable Air Traffic Management*. 2nd Edition. Brussels, Belgium: SESAR. Available at: https://www.atmmasterplan.eu/ [accessed: March 20, 2014].

Springett, D. 2005. Editorial: Critical perspectives on sustainable development. *Sustainable Development*, 13 (4), 209–11. Available at: http://dx.doi.org/10.1002/sd.279 (subscription required) [accessed: March 30, 2014].

Star Wars Episode V: The Empire Strikes Back. (film). 1980. Directed by I. Kershner. s.l.: Lucasfilm.

Steen, E.J. et al. 2010. Microbial production of fatty-acid-derived fuels and chemicals from plant biomass. *Nature*, 463 (7280), 559–63. Available at: http://www.nature.com/nature/journal/v463/n7280/pdf/nature08721.pdf (subscription required) [accessed: March 24, 2014].

Stern, N. 2006. *Stern Review: The Economics of Climate Change*. Cambridge, U.K.: Available at: http://webarchive.nationalarchives.gov.uk/20130129110402/http://www.hm-treasury.gov.uk/stern_review_report.htm (pre-publication version) [accessed: April 3, 2014].

Stoner, A.F. and Wankel, C. 2008. *Global Sustainability Initiatives: New Models and New Approaches*. Charlotte, NC: Information Age.

Stranges, A.N. 2014. *Fischer-Tropsch Archive*. [Online, January 16, 2008]. Emerging Fuels Technology and SGC Energia SA. Available at: http://www. fischer-tropsch.org [accessed: March 24, 2014].

Sustainable Aviation Fuel Users Group. 2014. *Our Commitment to Sustainable Options*. [Online]. SAFUG. Available at: http://www.safug.org/safug-pledge/ [accessed: April 17, 2014].

Sweetman, B. 1994. Airbus hits the road with A3XX. *Interavia Business and Technology*, 49 (583), 12. Available at: http://www.highbeam.com/ doc/1G1-16444324.html (subscription required) [accessed: March 20, 2014].

Taylor, J.W.R. and Munson, K.M. 1973. *History of Aviation*. London, U.K.: Octopus Books.

Toy Story. (film). 1995. Directed by J. Lasseter. Emeryville, CA: Pixar Animation Studios and Walt Disney Pictures.

Turner, C. 2011. *The Leap: How to Survive and Thrive in the Sustainable Economy*. Toronto, ON: Random House Canada.

Tyler, T. 2013. *Remarks of Tony Tyler at the Greener Skies Conference in Hong Kong*. [Online, February 26, 2013]. IATA. Available at: http://www.iata.org/ pressroom/speeches/Pages/2013-02-26-01.aspx [accessed: April 13, 2014].

United Nations. 1992. *United Nations Framework Convention on Climate Change*. s.l.: UN. Available at: http://unfccc.int/resource/docs/convkp/conveng.pdf [accessed: March 28, 2014].

United Nations. 1995. *Framework Convention on Climate Change: Report of the Conference of the Parties on its First Session; Berlin, 28 March to 7 April, 1995*. s.l.: UN. Available at: http://unfccc.int/resource/docs/cop1/07a01.pdf [accessed: August 14, 2014].

United Nations. 1998. *Kyoto Protocol to the United Nations Framework Convention on Climate Change*. s.l.: UN. Available at: http://unfccc.int/ resource/docs/convkp/kpeng.pdf [accessed: March 28, 2014].

United Nations General Assembly. 1983. *Process of Preparation of the Environmental Perspective to the Year 2000 and Beyond*. (A/RES/38/161, 19 December 1983). New York, NY: United Nations. Available at: http://www. un.org/ga/search/view_doc.asp?symbol=A/RES/38/161&Lang=E&Area=RE SOLUTION [accessed: April 1, 2014].

United States Department of Agriculture. 2013. *Agriculture, Navy Secretaries Promote U.S. Military Energy Independence with "Farm-to-Fleet"*. [Online, December 11, 2013]. USDA. Available at: http://www.usda.gov/wps/portal/ usda/usdamediafb?contentid=2013/12/0237.xml&printable=true&contentido nly=true [accessed: April 15, 2014].

United States Department of Energy. Advanced Research Projects Agency–Energy (ARPA-E). 2014. *Electrofuels: Microorganisms for Liquid Transportation Fuel*. [Online]. ARPA-E. Available at: http://arpa-e.energy.gov/?q=projects/ view-programs (choose "Electrofuels" from menu) [accessed: March 4, 2014].

United States Environmental Protection Agency. 2009. *Inventory of U.S. Greenhouse Gas Emissions and Sinks: 1990–2007*. Washington, DC: US EPA.

Available at: http://www.epa.gov/climatechange/emissions/usgginv_archive. html [accessed: March 17, 2014].

United States Environmental Protection Agency. 2013. *Renewable Fuel Standard (RFS)*. [Online, December 10, 2013]. US EPA. Available at: http://www.epa. gov/otaq/fuels/renewablefuels/ [accessed: April 22, 2014].

United States. Congress (109th). 2005. *Energy Policy Act of 2005*. (Public Law 109-58—August 8, 2005). Washington, DC: Government Printing Office. Available at: http://www.gpo.gov/fdsys/pkg/PLAW-109publ58/pdf/PLAW-109publ58.pdf [accessed: April 15, 2014].

United States. Congress (110th). 2007. *Energy Independence and Security Act of 2007*. (HR 6). Washington, DC: Government Printing Office. Available at: http://www.gpo.gov/fdsys/pkg/BILLS-110hr6enr/pdf/BILLS-110hr6enr.pdf [accessed: April 22, 2014].

Upham, P. 2003. *Towards Sustainable Aviation*. London, U.K. and Sterling, VA: Earthscan Publications.

Vasigh, B., Fleming, K. and Mackay, L. 2010. *Foundations of Airline Finance: Methodology and Practice*. Farnham, U.K. and Burlington, VT: Ashgate.

Vasigh, B., Fleming, K. and Tacker, T. 2008. *Introduction to Air Transport Economics: From Theory to Applications*. Aldershot, U.K. and Burlington, VT: Ashgate.

Vorrath, S. 2013. Czech follows Spain in deciding to tax output from solar power. *RenewEconomy.com.au*. [Online, September 18, 2013]. Available at: http:// reneweconomy.com.au/2013/czech-follows-spain-in-deciding-to-tax-output-from-solar-power-49694 [accessed: 1April 10, 2014].

Wallace, B. 2012. Sustainable infrastructure gains momentum with new certification body. *Climate Change Business Journal*, 5 (11/12), 1–4. Available at: http://www.sustainabilityprofessionals.org/system/files/Envision. CCBJReprint2013.pdf [accessed: April 6, 2014].

Warwick, G. 2007. Honeywell's UOP wins a US military biofuels contract. *Flightglobal*. [Online, July 3, 2007]. Available at: http://www.flightglobal. com/news/articles/honeywell39s-uop-wins-a-us-military-biofuels-contract-215286/ [accessed: May 4, 2014].

Weitzman, M.L. 2007. A Review of the Stern Review on the Economics of Climate Change. *Journal of Economic Literature*, 45 (3), 703–24. Available at: http://www.aeaweb.org/articles.php?doi=10.1257/jel.45.3.703 (subscription required) [accessed: April 3, 2014].

Whiteman, G., Hope, C. and Wadhams, P. 2013. Vast costs of Arctic change. *Nature*, 499, 401–3. Available at: http://dx.doi.org/10.1038/499401a (subscription required) [accessed: April 11, 2014].

Wilcox, L.J., Shine, K.P. and Hoskins, B.J. 2012. Radiative forcing due to aviation water vapour emissions. *Atmospheric Environment*, 63, 1–13. Available at: http://dx.doi.org/10.1016/j.atmosenv.2012.08.072 (subscription required) [accessed: March 17, 2014].

Wolveridge, P.E. 2000. Aviation turbine fuels, in *Modern Petroleum Technology: Volume 2, Downstream*. 6th Edition, edited by A.G. Lucas. Chichester, U.K.: John Wiley and Sons, 287–98.

World Commission on Environment and Development. 1987. *Report of the World Commission on Environment and Development: Our Common Future ["Brundtland Report"]*. Geneva, Switzerland: United Nations. Available at: http://www.un-documents.net/our-common-future.pdf [accessed: March 31, 2014].

World Hunger Education Service. 2013. 2013 world hunger and poverty facts and statistics. *Hunger Notes*. [Online, July 27, 2013]. Available at: http://www.worldhunger.org/articles/Learn/world%20hunger%20facts%202002.htm [accessed: April 1, 2014].

Wright, R. 2004. *Short History of Progress*. Toronto, ON: House of Anansi Press.

Index